MAINTENANCE FUNDAMENTALS

2nd Edition

MAINTENANCE FUNDAMENTALS
2nd Edition

R. Keith Mobley

ELSEVIER
BUTTERWORTH
HEINEMANN

AMSTERDAM • BOSTON • HEIDELBERG • LONDON • NEW YORK • OXFORD
PARIS • SAN DIEGO • SAN FRANCISCO • SINGAPORE • SYDNEY • TOKYO

Elsevier Butterworth–Heinemann
200 Wheeler Road, Burlington, MA 01803, USA
Linacre House, Jordan Hill, Oxford OX2 8DP, UK

Library of Congress Cataloging-in-Publication Data
Application submitted

British Library Cataloguing-in-Publication Data
A catalogue record for this book is available from the British Library.

ISBN-13: 978-0-7506-7798-1
ISBN-10: 0-7506-7798-8

For information on all Butterworth–Heinemann publications
visit our Web site at www.bh.com

Transferred to Digital Printing 2009

CONTENTS

1

IMPACT OF MAINTENANCE

Maintenance costs, as defined by normal plant accounting procedures, are normally a major portion of the total operating costs in most plants. Traditional maintenance costs (i.e., labor and material) in the United States have escalated at a tremendous rate over the past 10 years. In 1981, domestic plants spent more than $600 billion to maintain their critical plant systems. By 1991, the costs had increase to more than $800 billion, and they were projected to top $1.2 trillion by the year 2000. These evaluations indicate that on average, one third, or $250 billion, of all maintenance dollars are wasted through ineffective maintenance management methods. American industry cannot absorb the incredible level of inefficiency and hope to compete in the world market.

Because of the exorbitant nature of maintenance costs, they represent the greatest potential short-term improvement. Delays, product rejects, scheduled maintenance downtime, and traditional maintenance costs—such as labor, overtime, and repair parts—are generally the major contributors to abnormal maintenance costs within a plant.

The dominant reason for this ineffective management is the lack of factual data that quantify the actual need for repair or maintenance of plant machinery, equipment, and systems. Maintenance scheduling has been and in many instances still is predicated on statistical trend data or on the actual failure of plant equipment.

Until recently, middle and corporate level management have ignored the impact of the maintenance operation on product quality, production costs, and more importantly on bottom-line profit. The general opinion has been "maintenance is

1

a necessary evil" or "nothing can be done to improve maintenance costs." Perhaps these were true statements 10 or 20 years ago. However, the developments of microprocessor or computer-based instrumentation that can be used to monitor the operating condition of plant equipment, machinery, and systems have provided the means to manage the maintenance operation. They have provided the means to reduce or eliminate unnecessary repairs, prevent catastrophic machine failures, and reduce the negative impact of the maintenance operation on the profitability of manufacturing and production plants.

Maintenance Philosophies

Industrial and process plants typically utilize two types of maintenance management: (1) run-to-failure, or (2) preventive maintenance.

Run-to-Failure Management

The logic of run-to-failure management is simple and straightforward. When a machine breaks, fix it. This "if it ain't broke, don't fix it" method of maintaining plant machinery has been a major part of plant maintenance operations since the first manufacturing plant was built, and on the surface sounds reasonable. A plant using run-to-failure management does not spend any money on maintenance until a machine or system fails to operate. Run-to-failure is a reactive management technique that waits for machine or equipment failure before any maintenance action is taken. It is in truth a no-maintenance approach of management. It is also the most expensive method of maintenance management.

Few plants use a true run-to-failure management philosophy. In almost all instances, plants perform basic preventive tasks (i.e., lubrication, machine adjustments, and other adjustments) even in a run-to-failure environment. However, in this type of management, machines and other plant equipment are not rebuilt nor are any major repairs made until the equipment fails to operate.

The major expenses associated with this type of maintenance management are: (1) high spare parts inventory cost, (2) high overtime labor costs, (3) high machine downtime, and (4) low production availability. Since there is no attempt to anticipate maintenance requirements, a plant that uses true run-to-failure management must be able to react to all possible failures within the plant. This reactive method of management forces the maintenance department to maintain extensive spare parts inventories that include spare machines or at least all major components for all critical equipment in the plant. The alternative is to rely on equipment vendors that can provide immediate delivery of all required spare parts. Even if the latter is possible, premiums for expedited delivery substantially increase the costs of repair parts and downtime required for correcting machine failures. To minimize the impact on production created by unexpected machine failures, maintenance personnel must also be able to react immediately to all machine failures.

The net result of this reactive type of maintenance management is higher maintenance cost and lower availability of process machinery. Analysis of maintenance costs indicates that a repair performed in the reactive or run-to-failure mode will average about three times higher than the same repair made within a scheduled or preventive mode. Scheduling the repair provides the ability to minimize the repair time and associated labor costs. It also provides the means of reducing the negative impact of expedited shipments and lost production.

Preventive Maintenance Management

There are many definitions of preventive maintenance, but all preventive maintenance management programs are time driven. In other words, maintenance tasks are based on elapsed time or hours of operation. Figure 1.1 illustrates an example of the statistical life of a machine-train. The mean time to failure (MTTF) or bathtub curve indicates that a new machine has a high probability of failure, because of installation problems, during the first few weeks of operation. After this initial period, the probability of failure is relatively low for an extended period of time. Following this normal machine life period, the probability of failure increases sharply with elapsed time. In preventive maintenance management, machine repairs or rebuilds are scheduled on the basis of the MTTF statistic.

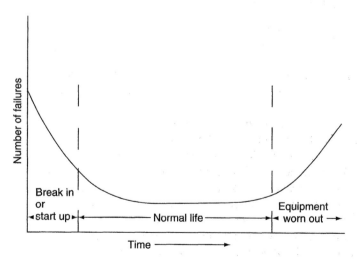

Figure 1.1 Bathtub curve.

The actual implementation of preventive maintenance varies greatly. Some programs are extremely limited and consist of lubrication and minor adjustments. More comprehensive preventive maintenance programs schedule repairs, lubrication, adjustments, and machine rebuilds for all critical machinery in the plant. The common denominator for all of these preventive maintenance programs is the scheduling guideline. All preventive maintenance management programs assume that machines will degrade within a time frame typical of its particular classification. For example, a single-stage, horizontal split-case centrifugal pump will normally run 18 months before it must be rebuilt. When preventive management techniques are used, the pump would be removed from service and rebuilt after 17 months of operation.

The problem with this approach is that the mode of operation and system or plant-specific variables directly affect the normal operating life of machinery. The mean time between failures (MTBF) will not be the same for a pump that is handling water and one that is handling abrasive slurries. The normal result of using MTBF statistics to schedule maintenance is either unnecessary repairs or catastrophic failure. In the example, the pump may not need to be rebuilt after 17 months. Therefore the labor and material used to make the repair was wasted. The second option, use of preventive maintenance, is even more costly. If the pump fails before 17 months, we are forced to repair by using run-to-failure techniques. Analysis of maintenance costs has shown that a repair made in a reactive mode (i.e., after failure) will normally be three times greater than the same repair made on a scheduled basis.

Predictive Maintenance

Like preventive maintenance, predictive maintenance has many definitions. To some, predictive maintenance is monitoring the vibration of rotating machinery in an attempt to detect incipient problems and to prevent catastrophic failure. To others, it is monitoring the infrared image of electrical switchgears, motors, and other electrical equipment to detect developing problems.

The common premise of predictive maintenance is that regular monitoring of the mechanical condition of machine-trains will ensure the maximum interval between repair and minimize the number and cost of unscheduled outages created by machine-train failures. Predictive maintenance is much more. It is the means of improving productivity, product quality, and overall effectiveness of our manufacturing and production plants. Predictive maintenance is not vibration monitoring or thermal imaging or lubricating oil analysis or any of the other nondestructive testing techniques that are being marketed as predictive maintenance tools. Predictive maintenance is a philosophy or attitude that, simply stated, uses the actual operating condition of plant equipment and

systems to optimize total plant operation. A comprehensive predictive maintenance management program utilizes a combination of the most cost-effective
tools—that is, vibration monitoring, thermography, tribology, etc.—to obtain
the actual operating condition of critical plant systems, and based on these actual
data, schedules all maintenance activities on an as-needed basis. Including
predictive maintenance in a comprehensive maintenance management program
will provide the ability to optimize the availability of process machinery and
greatly reduce the cost of maintenance. It will also provide the means to improve
product quality, productivity, and profitability of our manufacturing and
production plants.

Predictive maintenance is a condition-driven preventive maintenance program.
Instead of relying on industrial or in-plant average-life statistics (i.e., MTTF) to
schedule maintenance activities, predictive maintenance uses direct monitoring
of the mechanical condition, system efficiency, and other indicators to determine
the actual MTTF or loss of efficiency for each machine-train and system in the
plant. At best, traditional time-driven methods provide a guideline to normal
machine-train life spans. The final decision, in preventive or run-to-failure
programs, on repair or rebuild schedules must be made on the bases of intuition
and the personal experience of the maintenance manager. The addition of a
comprehensive predictive maintenance program can and will provide factual
data on the actual mechanical condition of each machine-train and operating
efficiency of each process system. These data provide the maintenance manager
with actual data for scheduling maintenance activities.

A predictive maintenance program can minimize unscheduled breakdowns of all
mechanical equipment in the plant and ensure that repaired equipment is in
acceptable mechanical condition. The program can also identify machine-train
problems before they become serious. Most mechanical problems can be minimized if they are detected and repaired early. Normal mechanical failure modes
degrade at a speed directly proportional to their severity. If the problem is
detected early, major repairs, in most instances, can be prevented. Simple vibration analysis is predicated on two basic facts: all common failure modes have
distinct vibration frequency components that can be isolated and identified, and
the amplitude of each distinct vibration component will remain constant unless
there is a change in the operating dynamics of the machine-train. These facts,
their impact on machinery, and methods that will identify and quantify the root
cause of failure modes will be developed in more detail in later chapters.

Predictive maintenance that utilizes process efficiency, heat loss, or other nondestructive techniques can quantify the operating efficiency of non-mechanical
plant equipment or systems. These techniques used in conjunction with vibration
analysis can provide the maintenance manager or plant engineer with factual

information that will enable him to achieve optimum reliability and availability from the plant.

There are five nondestructive techniques normally used for predictive maintenance management: (1) vibration monitoring, (2) process parameter monitoring, (3) thermography, (4) tribology, and (5) visual inspection. Each technique has a unique data set that will assist the maintenance manager in determining the actual need for maintenance. How do you determine which technique or techniques are required in your plant? How do you determine the best method to implement each of the technologies? If you listen to the salesman for the vendors that supply predictive maintenance systems, his is the only solution to your problem. How do you separate the good from the bad? Most comprehensive predictive maintenance programs will use vibration analysis as the primary tool. Since the majority of normal plant equipment is mechanical, vibration monitoring will provide the best tool for routine monitoring and identification of incipient problems. However, vibration analysis will not provide the data required on electrical equipment, areas of heat loss, condition of lubricating oil, or other parameters that should be included in your program.

ROLE OF MAINTENANCE ORGANIZATION

Too many maintenance organizations continue to pride themselves on how fast they can react to a catastrophic failure or production interruption rather than on their ability to prevent these interruptions. While few will admit their continued adherence to this breakdown mentality, most plants continue to operate in this mode. Contrary to popular belief, the role of the maintenance organization is to maintain plant equipment, not to repair it after a failure.

The mission of maintenance in a world-class organization is to achieve and sustain optimum availability.

Optimum Availability

The production capacity of a plant is, in part, determined by the availability of production systems and their auxiliary equipment. The primary function of the maintenance organization is to ensure that all machinery, equipment, and systems within the plant are always on line and in good operating condition.

Optimum Operating Condition

Availability of critical process machinery is not enough to ensure acceptable plant performance levels. The maintenance organization has the responsibility to maintain all direct and indirect manufacturing machinery, equipment, and systems so that they will be continuously in optimum operating condition. Minor problems, no matter how slight, can result in poor product quality, reduce production speeds, or affect other factors that limit overall plant performance.

Maximum Utilization of Maintenance Resources

The maintenance organization controls a substantial part of the total operating budget in most plants. In addition to an appreciable percentage of the total plant labor budget, the maintenance manager, in many cases, controls the spare parts inventory, authorizes the use of outside contract labor, and requisitions millions of dollars in repair parts or replacement equipment. Therefore, one goal of the maintenance organization should be the effective use of these resources.

Optimum Equipment Life

One way to reduce maintenance cost is to extend the useful life of plant equipment. The maintenance organization should implement programs that will increase the useful life of all plant assets.

Minimum Spares Inventory

Reductions in spares inventory should be a major objective of the maintenance organization. However, the reduction cannot impair their ability to meet goals 1 through 4. With the predictive maintenance technologies that are available today, maintenance can anticipate the need for specific equipment or parts far enough in advance to purchase them on an as-needed basis.

Ability to React Quickly

Not all catastrophic failures can be avoided. Therefore the maintenance organization must maintain the ability to react quickly to the unexpected failure.

EVALUATION OF THE MAINTENANCE ORGANIZATION

One means to quantify the maintenance philosophy in your plant is to analyze the maintenance tasks that have occurred over the past two to three years. Attention should be given to the indices that define management philosophy.

One of the best indices of management attitude and the effectiveness of the maintenance function is the number of production interruptions caused by maintenance-related problems. If production delays represent more than 30%

of total production hours, reactive or breakdown response is the dominant management philosophy. To be competitive in today's market, delays caused by maintenance-related problems should represent less than 1% of the total production hours.

Another indicator of management effectiveness is the amount of maintenance overtime required to maintain the plant. In a breakdown maintenance environment, overtime cost is a major negative cost. If your maintenance department's overtime represents more than 10% of the total labor budget, you definitely qualify as a breakdown operation. Some overtime is and will always be required. Special projects and the 1% of delays caused by machine failures will force some expenditure of overtime premiums, but these abnormal costs should be a small percentage of the total labor costs. Manpower utilization is another key to management effectiveness. Evaluate the percentage of maintenance labor as compared with total available labor hours that are expended on the actual repairs and maintenance prevention tasks. In reactive maintenance management, the percentage will be less than 50%. A well-managed maintenance organization should maintain consistent manpower utilization above 90%. In other words, at least 90% of the available maintenance labor hours should be effectively utilized to improve the reliability of critical plant systems, not waiting on something to break.

Three Types of Maintenance

There are three main types of maintenance and three major divisions of preventive maintenance, as illustrated in Figure 1.2.

Maintenance Improvement

Picture these divisions as the five fingers on your hand. Improvement maintenance efforts to reduce or eliminate the need for maintenance are like the thumb, the first and most valuable digit. We are often so involved in maintaining that we forget to plan and eliminate the need at its source. Reliability engineering efforts should emphasize elimination of failures that require maintenance. This is an opportunity to pre-act instead of react.

For example, many equipment failures occur at inboard bearings that are located in dark, dirty, inaccessible locations. The oiler does not lubricate inaccessible bearings as often as he lubricates those that are easy to reach. This is a natural tendency. One can consider reducing the need for lubrication by using permanently lubricated, long-life bearings. If that is not practical, at least an automatic oiler could be installed. A major selling point of new automobiles is the elimination of ignition points that require replacement and adjustment, introduction of self-adjusting brake shoes and clutches, and extension of oil change intervals.

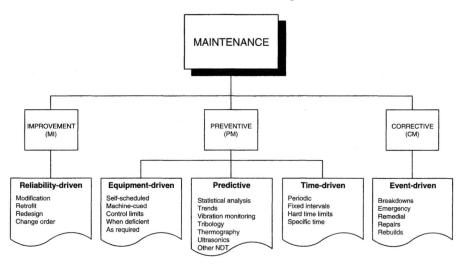

Figure 1.2 Structure of maintenance.

Corrective Maintenance

The little finger in our analogy to a human hand represents corrective maintenance (emergency, repair, remedial, unscheduled). At present, most maintenance is corrective. Repairs will always be needed. Better improvement maintenance and preventive maintenance, however, can reduce the need for emergency corrections. A shaft that is obviously broken into pieces is relatively easy to maintain because little human decision is involved. Troubleshooting and diagnostic fault detection and isolation are major time consumers in maintenance. When the problem is obvious, it can usually be corrected easily. Intermittent failures and hidden defects are more time consuming, but with diagnostics, the causes can be isolated and then corrected. From a preventive maintenance perspective, the problems and causes that result in failures provide the targets for elimination by viable preventive maintenance. The challenge is to detect incipient problems before they lead to total failures and to correct the defects at the lowest possible cost. That leads us to the middle three fingers the branches of preventive maintenance.

Preventive Maintenance

As the name implies, preventive maintenance tasks are intended to prevent un-scheduled downtime and premature equipment damage that would result in corrective or repair activities. This maintenance management approach is predom-inately a time-driven schedule or recurring tasks, such as lubrication and adjust-ments that are designed to maintain acceptable levels of reliability and availability.

Reactive Maintenance

Reactive maintenance is done when equipment needs it. Inspection with human senses or instrumentation is necessary, with thresholds established to indicate when potential problems start. Human decisions are required to establish those standards in advance so that inspection or automatic detection can determine when the threshold limit has been exceeded. Obviously, a relatively slow deterioration before failure is detectable by condition monitoring, whereas rapid, catastrophic modes of failure may not be detected. Great advances in electronics and sensor technology are being made.

Also needed is a change in the human thought process. Inspection and monitoring should include disassembly of equipment only when a problem is detected. The following are general rules for on-condition maintenance:

- Inspect critical components.
- Regard safety as paramount.
- Repair defects.
- If it works, don't fix it.

Condition Monitoring

Statistics and probability theory provide are the bases for condition monitor maintenance. Trend detection through data analysis often rewards the analyst with insight into the causes of failure and preventive actions that will help avoid future failures. For example, stadium lights burn out within a narrow range of time. If 10% of the lights have burned out, it may be accurately assumed that the rest will fail soon and should, most effectively, be replaced as a group rather than individually.

Scheduled Maintenance

Scheduled, fixed-interval preventive maintenance tasks should generally be used only if there is opportunity for reducing failures that cannot be detected in advance, or if dictated by production requirements. The distinction should be drawn between fixed-interval maintenance and fixed-interval inspection that may detect a threshold condition and initiate condition monitor tasks. Examples of fixed interval tasks include 3,000-mile oil changes and 48,000-mile spark plug changes on a car, whether it needs the changes or not. This may be very wasteful, because all equipment and their operating environments are not alike. What is right for one situation may not be right for another.

2

FUNDAMENTAL REQUIREMENTS OF EFFECTIVE PREVENTIVE MAINTENANCE

When most people think of preventive maintenance, they visualize scheduled, fixed-interval maintenance that is done every day, every month, every quarter, every season, or at some other predetermined interval. Timing may be based on days or on intervals such as miles, gallons, activations, or hours of use. The use of performance intervals is itself a step toward basing preventive tasks on actual need instead of just on a generality.

The two main elements of fixed-interval preventive maintenance are procedure and discipline. Procedure means that the correct tasks are done and the right lubricants applied and consumables replaced at the best interval. Discipline requires that all the tasks are planned and controlled so that everything is done when it should be done. Both of these areas deserve attention. The topic of procedures is covered in detail in the following sections.

Discipline is a major problem in many organizations. This is obvious when one considers the fact that many organizations do not have an established program. Further, organizations that do claim to have a program often fail to establish a good planning and control procedure to ensure accomplishment. Elements of such a procedure include:

1. Listing of all equipment and the intervals at which it must receive PMs
2. A master schedule for the year that breaks down tasks by month, week, and possibly even to the day

3. Assignment of responsible persons to do the work
4. Inspection by the responsible supervisor to make sure that quality work is done on time
5. Updating of records to show when the work was done and when the next preventive task is due
6. Follow-up as necessary to correct any discrepancies.

Fundamental Requirements of Effective Maintenance

Effective maintenance is not magic, nor is it dependent on exotic technologies or expensive instruments or systems. Instead, it is dependent on doing simple, basic tasks that will result in reliable plant systems. These basics include the following.

Inspections

Careful inspection, which can be done without "tearing down" the machine, saves both technician time and exposure of the equipment to possible damage. Rotating components find their own best relationship to surrounding components. For example, piston rings in an engine or compressor cylinder quickly wear to the cylinder wall configuration. If they are removed for inspection, the chances are that they will not easily fit back into the same pattern. As a result, additional wear will occur and the rings will have to be replaced much sooner than if they were left intact and performance-tested for pressure produced and metal particles in the lubricating oil.

Human Senses

We humans have a great capability for sensing unusual sights, sounds, smells, tastes, vibrations, and touches. Every maintenance manager should make a concerted effort to increase the sensitivity of his own and that of his personnel's human senses. Experience is generally the best teacher. Often, however, we experience things without knowing what we are experiencing. A few hours of training in what to look for could have high payoff.

Human senses are able to detect large differences but are generally not sensitive to small changes. Time tends to have a dulling effect. Have you ever tried to determine if one color is the same as another without having a sample of each to compare side by side? If you have, you will understand the need for standards. A standard is any example that can be compared with the existing situation as a measurement. Quantitative specifications, photographs, recordings, and actual samples should be provided. The critical parameters should be clearly marked on them with a display as to what is good and what is bad.

As the reliability-based preventive maintenance program develops, samples should be collected that will help to pinpoint with maximum accuracy how much wear can take place before problems will occur. A display where craftsmen gather can be effective. A framed 4-foot by 4-foot pegboard works well since shafts, bearings, gears, and other components can be easily wired to it or hung on hooks for display. An effective but little-used display area where notices can be posted is above the urinal or on the inside of the toilet stall door. Those are frequently viewed locations and allow people to make dual use of their time.

Sensors

Since humans are not continually alert or sensitive to small changes and cannot get inside small spaces, especially when machines are operating, it is necessary to use sensors that will measure conditions and transmit information to external indicators.

Sensor technology is progressing rapidly; there have been considerable improvements in capability, accuracy, size, and cost. Pressure transducers, temperature thermocouples, electrical ammeters, revolution counters, and a liquid height level float are examples found in most automobiles.

Accelerometers, eddy-current proximity sensors, and velocity seismic transducers are enabling the techniques of motion, position, and expansion analysis to be increasingly applied to large numbers of rotating machines. Motors, turbines, compressors, jet engines, and generators can use vibration analysis. The normal pattern of operation, called its "signature," is established by measuring the performance of equipment under known good conditions. Comparisons are made at routine intervals, such as every 30 days, to determine if any of the parameters are changing erratically, and further, what the effect of such changes may be.

Spectrometric oil analysis process is useful for any mechanical moving device that uses oil for lubrication. It tests for the presence of metals, water, glycol, fuel dilution, viscosity, and solid particles. Automotive engines, compressors, and turbines all benefit from oil analysis. Most major oil companies will provide this service if you purchase lubricants from them.

The major advantage of spectrometric oil analysis is early detection of component wear. Not only does it evaluate when oil is no longer lubricating properly and should be replaced, it also identifies and measures small quantities of metals that are wearing from the moving surfaces. The metallic elements found, and their quantity, can indicate which components are wearing and to what degree so that maintenance and overhaul can be carefully planned. For example, the presence of chrome would indicate cylinder-head wear, phosphor bronze would probably be from the main bearings, and stainless steel would point toward lifters. Experience with particular equipment naturally leads to improved diagnosis.

Thresholds

Now that instrumentation is becoming available to measure equipment performance, it is still necessary to determine when that performance is "go" and when it is "no go." A human must establish the threshold point, which can then be controlled by manual, semi-automatic, or automatic means. First, let's decide how the threshold is set and then discuss how to control it.

To set the threshold, one must gather information on what measurements can exist while equipment is running safely and what the measurements are just prior to or at the time of failure. Equipment manufacturers, and especially their experienced field representatives, will be a good starting source of information. Most manufacturers will run equipment until failure in their laboratories as part of their tests to evaluate quality, reliability, maintainability, and maintenance procedures. Such data are necessary to determine under actual operating conditions how much stress can be put on a device before it will break. Many devices that should not be taken to the breaking point under operating conditions, such as nuclear reactors and flying airplanes, can be made to fail under secure test conditions so that knowledge can be used to keep them safe during actual use.

Once the breaking point is determined, a margin of safety should be added to account for variations in individual components, environments, and operating conditions. Depending on the severity of failure, that safety margin could be anywhere from one to three standard deviations before the average failure point. One standard deviation on each side of the mean will include 68% of all variations, two standard deviations will include 95%, and three standard deviations will include 98.7%. Where our mission is to prevent failures, however, only the left half of the distribution is applicable. This single-sided distribution also shows that we are dealing with probabilities and risk.

The earlier the threshold is set and effective preventive maintenance done, the greater is the assurance that it will be done prior to failure. If the MTBF is 9,000 miles with a standard deviation of 1,750 miles, then proper preventive maintenance at 5,500 miles could eliminate almost 98% of the failures. Note the word "proper," meaning that no new problems are injected. That also means, however, that costs will be higher than need be since components will be replaced before the end of their useful life, and more labor will be required.

Once the threshold set point has been determined, it should be monitored to detect when it is exceeded. The investment in monitoring depends on the period over which deterioration may occur, the means of detection, and the benefit value. If failure conditions build up quickly, a human may not easily detect the condition, and the relatively high cost of automatic instrumentation will be repaid.

Lubrication

The friction of two materials moving relative to each other causes heat and wear. Friction-related problems cost industries over $1 billion per annum. Technology intended to improve wear resistance of metal, plastics, and other surfaces in motion has greatly improved over recent years, but planning, scheduling, and control of the lubricating program is often reminiscent of a plant handyman wandering around with his long-spouted oil can.

Anything that is introduced onto or between moving surfaces to reduce friction is called a *lubricant*. Oils and greases are the most commonly used substances, although many other materials may be suitable. Other liquids and even gases are being used as lubricants. Air bearings, for example, are used in gyroscopes and other sensitive devices in which friction must be minimal. The functions of a lubricant are to:

1. Separate moving materials from each other to prevent wear, scoring, and seizure
2. Reduce heat
3. Keep out contaminants
4. Protect against corrosion
5. Wash away worn materials.

Good lubrication requires two conditions: sound technical design for lubrication and a management program to ensure that every item of equipment is properly lubricated.

Lubrication Program Development

Information for developing lubrication specifications can come from four main sources:

1. Equipment manufacturers
2. Lubricant vendors
3. Other equipment users
4. Individuals' own experience.

As with most other preventive maintenance elements, initial guidance on lubrication should come from manufacturers. They should have extensive experience with their own equipment, both in their test laboratories and in customer locations. They should know what parts wear and are frequently replaced. Therein lies a caution—a manufacturer could in fact make short-term profits by selling large numbers of spare parts to replace worn ones. Over the long term, however, that strategy will backfire and other vendors, whose equipment is less prone to wear and failure, will replace them.

Lubricant suppliers can be a valuable source of information. Most major oil companies will invest considerable time and effort in evaluating their customers' equipment to select the best lubricants and frequency or intervals for change. Naturally, these vendors hope that the consumer will purchase their lubricants, but the result can be beneficial to everyone. Lubricant vendors perform a valuable service of communicating and applying knowledge gained from many users to their customers' specific problems and opportunities.

Experience gained under similar operating conditions by other users or in your own facilities can be one of the best teachers. Personnel, including operators and mechanics, have a major impact on lubrication programs.

A major step in developing the lubrication program is to assign specific responsibility and authority for the lubrication program to a competent maintainability or maintenance engineer. The primary functions and steps involved in developing the program are to

1. Identify every piece of equipment that requires lubrication
2. Ensure that every major piece of equipment is uniquely identified, preferably with a prominently displayed number
3. Ensure that equipment records are complete for manufacturer and physical location
4. Determine locations on each piece of equipment that need to be lubricated
5. Identify lubricant to be used
6. Determine the best method of application
7. Establish the frequency or interval of lubrication
8. Determine if the equipment can be safely lubricated while operating or if it must be shut down
9. Decide who should be responsible for any human involvement
10. Standardize lubrication methods
11. Package the above elements into a lubrication program
12. Establish storage and handling procedures

13. Evaluate new lubricants to take advantage of state of the art
14. Analyze any failures involving lubrication and initiate necessary corrective actions.

An individual supervisor in the maintenance department should be assigned the responsibility for implementation and continued operation of the lubrication program. This person's primary functions are to

1. Establish lubrication service actions and schedules
2. Define the lubrication routes by building, area, and organization
3. Assign responsibilities to specific persons
4. Train lubricators
5. Ensure supplies of proper lubricants through the storeroom
6. Establish feedback that ensures completion of assigned lubrication and follows up on any discrepancies
7. Develop a manual or computerized lubrication scheduling and control system as part of the larger maintenance management program
8. Motivate lubrication personnel to check equipment for other problems and to create work requests where feasible
9. Ensure continued operation of the lubrication system.

It is important that a responsible person who recognizes the value of thorough lubrication be placed in charge. As with any activity, interest diminishes over time, equipment is modified without corresponding changes to the lubrication procedures, and state-of-the-art advances in lubricating technology may not be undertaken. A factory may have thousands of lubricating points that require attention. Lubrication is no less important to computer systems even though they are often perceived as electronic. The computer field engineer must provide proper lubrication to printers, tape drives, and disks that spin at 3,600 rpm. A lot of maintenance time is invested in lubrication. The effect on production uptime can be measured nationally in billions of dollars.

Calibration

Calibration is a special form of preventive maintenance whose objective is to keep measurement and control instruments within specified limits. A "standard" must be used to calibrate the equipment. Standards are derived from parameters established by the National Bureau of Standards (NBS). Secondary standards that have been manufactured to close tolerances and set against the primary standard are available through many test and calibration laboratories and often in industrial and university tool rooms and research labs. Ohmmeters are examples of equipment that should be calibrated at least once a year and before further use if subjected to sudden shock or stress.

The government sets forth calibration system requirements in MIL-C-45662 and provides a good outline in the military standardization handbook MIL-HDBK-52, *Evaluation of Contractor's Calibration System.* The principles are equally applicable to any industrial or commercial situation. The purpose of a calibration system is to provide for the prevention of tool inaccuracy through prompt detection of deficiencies and timely application of corrective action. Every organization should prepare a written description of its calibration system. This description should cover the measuring of test equipment and standards, including the following:

1. Establishment of realistic calibration intervals
2. List of all measurement standards
3. Established environmental conditions for calibration
4. Ensuring the use of calibration procedures for all equipment and standards
5. Coordinating the calibration system with all users
6. Ensuring that equipment is frequently checked by periodic system or cross-checks to detect damage, inoperative instruments, erratic readings, and other performance-degrading factors that cannot be anticipated or provided for by calibration intervals
7. Provide for timely and positive correction action
8. Establish decals, reject tags, and records for calibration labeling
9. Maintain formal records to ensure proper controls.

The checking interval may be in terms of time (hourly, weekly, monthly) or based on amount of use (e.g., every 5,000 parts, or every lot). For electrical test equipment, the power-on time may be a critical factor and can be measured through an electrical elapsed-time indicator.

Adherence to the checking schedule makes or breaks the system. The interval should be based on stability, purpose, and degree of usage. If initial records indicate that the equipment remains within the required accuracy for successive calibrations, then the intervals may be lengthened. On the other hand, if equipment requires frequent adjustment or repair, the intervals should be shortened. Any equipment that does not have specific calibration intervals should be (1) examined at least every 6 months and (2) calibrated at intervals of no longer than 1 year.

Adjustments or assignment of calibration intervals should be done in such a way that a minimum of 95% of equipment or standards of the same type is within tolerance when submitted for regularly scheduled recalibration. In other words, if more than 5% of a particular type of equipment is out of tolerance at the end of its interval, then the interval should be reduced until less than 5% is defective when checked.

A record system should be kept on every instrument, including the following:

1. History of use
2. Accuracy
3. Present location
4. Calibration interval and when due
5. Calibration procedures and necessary controls
6. Actual values of latest calibration
7. History of maintenance and repairs.

Test equipment and measurement standards should be labeled to indicate the date of last calibration, by whom it was calibrated, and when the next calibration is due. When the size of the equipment limits the application of labels, an identifying code should be applied to reflect the serviceability and due date for next calibration. This provides a visual indication of the calibration serviceability status. Both the headquarters calibration organization and the instrument user should maintain a two-way check on calibration. A simple means of doing this is to have a small form for each instrument with a calendar of weeks or months (depending on the interval required) across the top that can be punched and noted to indicate the calibration due date.

Planning and Estimating

Planning is the heart of good inspection and preventive maintenance. As described earlier, the first thing to establish is what items must be maintained and what the best procedure is for performing that task. Establishing good procedures requires a good deal of time and talent. This can be a good activity for a new graduate engineer, perhaps as part of a training process that rotates him or her through various disciplines in a plant or field organization. This experience can be excellent training for a future design engineer.

Writing ability is an important qualification, along with pragmatic experience in maintenance practices. The language used should be clear and concise, with short sentences. Who, what, when, where, why, and how should be clearly described. The following points should be noted from this typical procedure:

1. Every procedure has an identifying number and title.
2. The purpose is outlined.
3. Tools, reference documents, and any parts are listed.

   4. Safety and operating cautions are prominently displayed.
   5. A location is clearly provided for the maintenance mechanic to indicate performance as either satisfactory or deficient. If it is deficient, details are written in the space provided at the bottom for planning further work.

The procedure may be printed on a reusable, plastic-covered card that can be pulled from the file, marked, and returned when the work order is complete, on a standard preprinted form, or on a form that is uniquely printed by computer each time a related work order is prepared.

Whatever the medium of the form, it should be given to the preventive maintenance craftsperson together with the work order so that he or she has all the necessary information at his or her fingertips. The computer version has the advantage of single-point control that may be uniformly distributed to many locations. This makes it easy for an engineer at headquarters to prepare a new procedure or to make any changes directly on the computer and have them instantly available to any user in the latest version.

Two slightly different philosophies exist for accomplishing the unscheduled actions that are necessary to repair defects found during inspection and preventive maintenance. One is to fix them on the spot. The other is to identify them clearly for later corrective action. This logic was outlined in Figure 1.2. If a "priority one" defect that could hurt a person or cause severe damage is observed, the equipment should be immediately stopped and "red tagged" so that it will not be used until repairs are made. Maintenance management should establish a guideline such as, "Fix anything that can be corrected within 10 minutes, but if it will take longer, write a separate work request." The policy time limit should be set, based on

   1. Travel time to that work location
   2. Effect on production
   3. Need to keep the craftsperson on a precise time schedule.

The inspector who finds them can effect many small repairs most quickly. This avoids the need for someone else to travel to that location, identify the problem, and correct it. And it provides immediate customer satisfaction. More time-consuming repairs would disrupt the inspector's plans, which could cause other, even more serious problems to go undetected. The inspector is like a general practitioner who performs a physical exam and may give advice on proper diet and exercise but who refers any problems he may find to a specialist.

The inspection or preventive maintenance procedure form should have space where any additional action required can be indicated. When the procedure is completed and turned into maintenance control, the planner or scheduler should note any additional work required and see that it gets done according to priority.

Estimating Time

Since inspection or preventive maintenance is a standardized procedure with little variation, the tasks and time required can be accurately estimated. Methods of developing time estimates include consideration of such resources as

1. Equipment manufacturers' recommendations
2. National standards such as Chilton's on automotive or Means' for facilities
3. Industrial engineering time-and-motion studies
4. Historical experience.

Experience is the best teacher, but it must be carefully critiqued to make sure that the "one best way" is being used and that the pace of work is reasonable.

The challenge in estimating is to plan a large percentage of the work (preferably at least 90%) so that the time constraints are challenging but achievable without a compromise in high quality. The tradeoff between reasonable time and quality requires continuous surveillance by experienced supervisors. Naturally, if a maintenance mechanic knows that his work is being time studied, he will follow every procedure specifically and will methodically check off each step of the procedure. When the industrial engineer goes away, the mechanic will do what he feels are necessary items, in an order that may or may not be satisfactory. As is discussed earlier, regarding motivation, an experienced preventive maintenance inspector mechanic can vary performance as much as 50% either way from standard without most maintenance supervisors recognizing a problem or opportunity for improvement. Periodic checking against national time-and-motion standards, as well as trend analysis of repetitive tasks, will help keep preventive task times at a high level of effectiveness.

Estimating Labor Cost

Cost estimates follow from time estimates simply by multiplying the hours required by the required labor rates. Beware of coordination problems in which multiple crafts are involved. For example, one "Fortune 100" company has trade jurisdictions that require the following personnel in order to remove an electric motor: a tinsmith to remove the cover, an electrician to disconnect the electrical supply, a millwright to unbolt the mounts, and one or more laborers to remove the motor from its mount. That situation is fraught with inefficiency and high labor costs, since all four trades must be scheduled together, with at least three people watching while the fourth is at work. The cost will be at least four times what it could be and is often greater if one of the trades does not show up on time. The best a scheduler can hope for is, if he has the latitude, to schedule the cover removal at, say, 8:00 a.m. and the other functions at reasonable time intervals thereafter: electrician at 9:00, millwright at 10:00, and laborers at 11:00.

It is recommended that estimates be prepared on "pure" time. In other words, the exact hours and minutes that would be required under perfect scheduling conditions should be used. Likewise, it should be assumed that equipment would be available from production immediately. Delay time should be reported, and scheduling problems should be identified so that they can be addressed separately from the hands-on procedure times. Note that people think in hours and minutes, so 1 hour and 10 minutes is easier to understand than 1.17 hours.

Estimating Materials

Most parts and materials that are used for preventive maintenance are well known and can be identified in advance. The quantity of each item planned should be multiplied by the cost of the item in inventory. The sum of those extended costs will be the material cost estimate. Consumables such as transmission oil should be enumerated as direct costs, but grease and other supplies used from bulk should be included in overhead costs.

Scheduling

Scheduling is, of course, one of the advantages of doing preventive maintenance over waiting until equipment fails and then doing emergency repairs. Like many other activities, the watchword should be "PADA," which stands for "Plan-a-Day-Ahead." In fact, the planning for inspections and preventive activities can be done days, weeks, and even months in advance to ensure that the most convenient time for production is chosen, that maintenance parts and materials are available, and that the maintenance workload is relatively uniform.

Scheduling is primarily concerned with balancing demand and supply. Demand comes from the equipment's need for preventive maintenance. Supply is the availability of the equipment, craftspeople, and materials that are necessary to do the work. Establishing the demand has been partially covered in the chapters on on-condition, condition monitoring, and fixed-interval preventive maintenance tasks. Those techniques identify individual equipment as candidates for preventive maintenance.

Coordination with Production

Equipment is not always available for preventive maintenance just when the maintenance schedulers would like it to be. An overriding influence on coordination should be a cooperative attitude between production and maintenance. This is best achieved by a meeting between the maintenance manager and production management, including the foreman level, so that what will be done to prevent failures, how this will be accomplished, and what production should expect to gain in uptime may all be explained.

The cooperation of the individual machine operators is of prime importance. They are on the spot and most able to detect unusual events that may indicate equipment malfunctions. Once an attitude of general cooperation is established, coordination should be refined to monthly, weekly, daily, and possibly even hourly schedules. Major shutdowns and holidays should be carefully planned so that any work that requires "cold" shutdown can be done during those periods. Maintenance will often find that they must do this kind of work on weekends and holidays, when other persons are on vacation. Normal maintenance should be coordinated according to the following considerations:

1. Maintenance should publish a list of all equipment that is needed for inspections, preventive maintenance, and modifications and the amount of cycle time that such equipment will be required from production.
2. A maintenance planner should negotiate the schedule with production planning so that a balanced workload is available each week.
3. By Wednesday of each week, the schedule for the following week should be negotiated and posted where it is available to all concerned; it should be broken down by days.
4. By the end of the day before the preventive activity is scheduled, the maintenance person who will do the preventive maintenance should have seen the first-line production supervisor in charge of the equipment to establish a specific time for the preventive task.
5. The craftsperson should make every effort to do the job according to schedule.
6. As soon as the work is complete, the maintenance person should notify the production supervisor so that the equipment may be put back into use.

Overdue work should be tracked and brought up to date. Preventive maintenance scheduling should make sure that the interval is maintained between preventive actions. For example, if a preventive task for May is done on the 30th of the month, the next monthly task should be done during the last week of June. It is foolish to do a preventive maintenance task on May 30 and another June 1 just to be able to say one was done each month. In the case of preventive maintenance, the important thing is not the score but how the game was played.

Ensuring Completion

A formal record is desirable for every inspection and preventive maintenance job. If the work is at all detailed, a checklist should be used. The completed checklist should be returned to the maintenance office on completion of the work. Any open preventive maintenance work orders should be kept on report until the supervisor has checked the results for quality assurance and signed off approval. Modern computer technology with handheld computers and pen-based electronic assistants permit paperless checklists and verification. In many situations, a paper work order form is still the most practical medium for the field technician. The collected data should then be entered into a computer system for tracking.

Record Keeping

The foundation records for preventive maintenance are the equipment files. In a small operation with less than 200 pieces of complex equipment, the records can easily be maintained on paper. The equipment records provide information for purposes other than preventive maintenance. The essential items include the following:

- Equipment identification number
- Equipment name
- Equipment product/group/class
- Location
- Use meter reading
- Preventive maintenance interval(s)
- Use per day
- Last preventive maintenance due
- Next preventive maintenance due
- Cycle time for preventive maintenance
- Crafts required, number of persons, and time for each
- Parts required.

Back to Basics

Obviously, effective maintenance management requires much more than these fundamental tasks. However, these basic tasks must be the foundation of every successful maintenance program. Other tools, such as CMMS, predictive maintenance, etc., cannot replace them.

3

DESIGNING A PREVENTIVE
MAINTENANCE PROGRAM

Valid failure data provide the intelligence for an effective preventive maintenance program. After all, the objective is to prevent those failures from recurring.
A failure reporting system should identify the problem, cause, and corrective
action for every call. An action group, prophetically called the Failure Review
and Corrective Actions Task Force (FRACAS), can be very effective for involving responsible organizations in both detailed identification of problems and
causes and assignment of both short- and long-term corrective action. The
following are typical factory and field problems and codes that shorten the
computer data entry to four or fewer characters:

NOOP	Not Operable	OTHR	Other
BELR	Below rate	PM	Preventive task
INTR	Intermittent	QUAL	Quality
LEAK	Leak	SAFE	Safety
MOD	Modification	WEAT	Weather
NOIS	Noise	NPF	No problem found

The following are typical cause codes:

1.	Not applicable	30.	Motors
10.	Controls	40.	Drivers
20.	Power	50.	Transports
21.	External input power	60.	Program
22.	Main power supply	70.	Materials

71.	Normal wear	90.	Environment
72.	Damaged	99.	No cause found
80.	Operator	PM.	Preventive maintenance

The typical action codes are as follows:

A/A	Adjust/align	REF	Refurbish
CAL	Calibrate	REB	Rebuild
CONS	Consumables	LUBE	Lubricate
DIAG	Diagnose	MOD	Modify
REMV	Remove	PM	Preventive task
R/R	Remove and replace	RPR	Repair
R/RE	Remove and reinstall	TRN	Train
INST	Install	NC	Not complete
INSP	Inspect	NK	Not known

These parameters and their codes should be established to fit the needs of the specific organization. For example, an organization with many pneumatic and optical instruments would have sticky dials and dirty optics that would not concern an electronically oriented organization. Note also that the code letters are the same, whenever possible, as the commonly used words' first letters. Preventive maintenance activities are recorded simply as PM. The cause codes, which may be more detailed, can use numbers and subsets of major groups, such as all power will be 20s, with external input power = 21, main power supply = 22, and so on.

It is possible, of course, to write out the complete words. However, analysis, whether done by computer or manually, requires standard terms. Short letter and number codes strike a balance that aids short reports and rapid data entry.

Use of the equipment at every failure should also be recorded. A key to condition monitoring preventive maintenance effectiveness is knowing how many hours, miles, gallons, activations, or other kinds of use have occurred before an item failed. This requires hour meters and similar instrumentation on major equipment. Use on related equipment may often be determined by its relationship to the parent. For example, it may be determined that if a specific production line is operating for 7 hours, then the input feeder operates 5 hours (5/7), the mixer 2 hours (2/7), and the packaging machine 4 hours (4/7).

It is also important to determine the valid relationship between the cause of the problem and the recording measurement. For example, failures of an automotive starter are directly related to the number of times the car engine is started and only indirectly to odometer miles. If startup or a particular activity stresses the equipment differently from normal use, those special activities should be recorded.

Figure 3.1 is a combination work order and completion form. This form is printed by the computer on plain paper with the details of the work order on the top, space in the center for labor and materials for work orders that take a day or less, and a completion blank at the bottom to show when the work was started, when it was completed, the problem/cause/action codes, and meter reading. Labor on work orders that take more than one day is added daily from time reports and accumulated against the work order. Figure 3.2 shows the computer input screen for a simple service call report form that gathers minimum information necessary for field reporting. Those forms may be used as input for a computer system, when a direct-entry system is not available.

IMPROVING EQUIPMENT RELIABILITY

Total Plant Performance Management (TPPM) and similar quality programs promote a holistic approach that includes equipment performance as a major enhancement to productivity. To reinforce the "five-fingered approach to effective maintenance" outlined in Chapter 1, the fundamental thumb is elimination of failures. Uptime of equipment is what counts.

Maintainability and maintenance are most successful if we don't have failures to fix. Successful maintenance organizations spend more time on identification of trends and eliminating problems than they spend fixing repetitive breakdowns. Computerized maintenance management systems provide a tool to gather data and provide analysis that can lead to improvement.

Improvement Process

Figure 3.3 diagrams a business improvement process. A maintenance organization should start by measuring its own performance. For example, just a breakout of a typical day in the life of a maintenance person will be revealing. Many groups are chagrined to discover that maintenance staff actually work less than 30% of the time. Benchmark comparisons with similar organizations provide a basis for analyzing performance both on metrics and processes. The third step in goal setting is to identify realistic ideal levels of performance. These goals should have the following characteristics:

- Written
- Measurable
- Understandable
- Challenging
- Achievable

Work Order				
ORDER #:1926	PAD#: 45524	TYPE: A		PRI: 9
REQUESTED BY: Joe Jones	DEPARTMENT Maint. Planning	TELEPHONE# EXT. 456	TGT START 5/30/00	TGT COMPLETE 12/23/03

DESCRIPTION		EQUIPMENT	
PM-A Recharge Freon in A/C 44		ID:	44
		NAME:	Air Conditioner
		LOC:	CNTR RM 16

SPECIAL EQUIPMENT		ASSIGNED EMPLOYEE		PRECAUTIONS	
Charger Kit		657890	ID:	PRD-PROD PERMT	
		Jones, Joe	NAME:		
DOC: A/C 544		ACCOUNTING:	453–789	100%	

LABOR USED (ONLY FOR SINGLE-DAY JOBS)

DATE:	PERSON OR EQUIPMENT		TOTAL HOURS-MINUTES			
		WORK	TVL	DELAY	OT	$

DATE:	MATERIAL POSTING				
	PART#	DESCRIPTION	QTY.	$ UNIT	$ TOTAL
9/23/03	603552	Freon, A/C Charge Kit	1	$12.75	$12.75

TOTAL MATERIAL COST:					$12.75

COMPLETION

	DATE	TIME		CODES:	CURR
STARTED:				PBM:	METER
COMPLETED:				CAU:	READ:
				ACT:	
SIGNATURE:			DATE:		

Figure 3.1 Combination work order and completion form.

SERVICE CALL

Call Number:	2521	Cust. Acct Nbr. 5492		
Employee Number:	2297	Name: Joe Smith	Facility Name: XYZ Compant	

Status: (ABC=1) (SYZ=2) (CNT=3)   Received: 05/03/2004   Complete: 06/03/2004

Equipment: C90-0001   Description: Ingersoll-Rant Compressor

Description: Replaced worn 1st stage pinion geart

Part Numbers	Description	Unit Costs	Quantity	Extended Cost	Codes:		
751133	Gear, pinion, 1st stage	1,190.00	1	1,190.00	PBM	CAU	ACT
100012	Gasket, case, 1st stage	180.00	1	180.00	MOD	40	MOD

Hours - Minutes

Work	Travel	Delay	Overtime
9-51	1-38	0-58	0-00

Other Equipment Worked On?   N

Total Call:	Hours	Labor	Materials	Total
	11-47	252.34	1,370.00	1,622.34

Figure 3.2 Computer input screen for a service call form, which gathers minimum information necessary for field reporting.

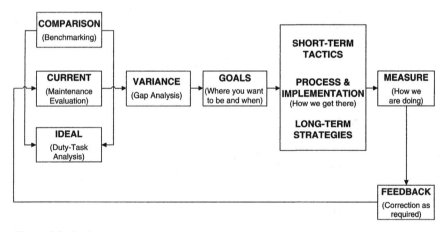

Figure 3.3 Business improvement process.

The goals will have firm times, dollars, percents, and dates. Everyone who will be challenged to meet the goals should be involved in their establishment. This may seem like a bureaucratic, warm-fuzzy approach, but the time it takes to achieve buy-in is earned back many times during accomplishment. Once the goals are set, any gaps between where performance is now versus where it needs to be can be identified. Then both short-term plans and long-term strategies can be implemented to reach the goals. Frequent measurement and feedback will revise performance to achieve the desired levels of achievement.

Failures That Can Be Prevented

Simplified Failure Modes and Effects Analysis (SFMEA) provides a method for determining which failures can be prevented. Necessary inputs are the frequency of occurrence for each problem and cause combination and what happens if a failure occurs. Criticality of the failure is considered for establishing priority of effort. SFMEA is a top-down approach that looks at major components in the equipment and asks, "Will it fail?" And if so, how and why? Preventive maintenance investigators are, of course, interested in how a component will fail so that the mechanism for failure can be reduced or eliminated. For example, heat is the most common cause of failure for electrical and mechanical components. Friction causes heat in assemblies moving relative to each other, often accompanied by material wear, and leads to many failures.

Any moving component is likely to fail at a relatively high rate and is a fine candidate for preventive maintenance. The following are familiar causes of failure:

- Abrasion
- Abuse
- Age deterioration
- Bond separation
- Consumable depletion
- Contamination
- Corrosion
- Dirt
- Fatigue
- Friction
- Operator negligence
- Puncture
- Shock
- Stress
- Temperature extremes
- Vibration
- Wear.

Maintenance To Prevent Failures

Cleanliness is the watchword of preventive maintenance. Metal filings, fluids in the wrong places, ozone and other gases that deteriorate rubber components—all are capable of damaging equipment and causing it to fail. A machine shop, for example, that contains many electro-mechanical lathes, mills, grinders, and boring machines should have established procedures for ensuring that the equipment is frequently cleaned and properly lubricated. In most plants, the best tactic is to assign responsibility for cleaning and lubrication to the machine's operator. There should be proper lubricants in grease guns and oil cans and cleaning materials at every workstation. Every operator should be trained in proper operator preventive tasks. A checklist should be kept on the equipment for the operator to initial every time the lubrication is done.

It is especially important that the lubrication be done cleanly. Grease fittings, for example, should be cleaned with waste material both before and after the grease gun is used. Grease attracts and holds particles of dirt. If the fittings are not clean, the grease gun could force contaminants between the moving parts, which is precisely what should be avoided. This is one example of how preventive maintenance done badly can be worse than no maintenance at all.

Personnel

Another tactic for ensuring thorough lubrication is to have an "oiler" who can do all of the lubrication at the beginning of each shift. This may be better than having the operators do lubrication if the task is at all complicated or if the operators are not sufficiently skilled.

Whether operators will do their own lubrication, rather than have it done by an oiler, is determined by

1. The complexity of the task
2. The motivation and ability of the operator
3. The extent of pending failures that might be detected by the oiler but overlooked by operators.

If operators can properly do the lubrication, then it should be made a part of their total responsibility, just as any car driver will make sure that he has adequate gasoline in his vehicle. It is best if the operators are capable of doing their own preventive maintenance. Like many tasks, preventive maintenance should be delegated to the lowest possible level consistent with adequate knowledge and ability. If, however, there is a large risk that operators may cause damage through negligence, willful neglect, or lack of ability, then a maintenance specialist should do lubrication. The tasks should be clearly defined. Operators may be able to do some items, while maintenance personnel will be required for others. Examples of how the work can be packaged will be described later.

Preventive tasks are often assigned to the newest maintenance trainee. In most cases, management is just asking for trouble if it is regarded as low-status, undesirable work. If management believes in preventive maintenance, they should assign well-qualified personnel. Education and experience make a big difference in maintenance. Most organizations have at least one skilled maintenance person who can simply step onto the factory floor and sense—through sight, sound, smell, vibration, and temperature—the conditions in the factory. This person can tell in an instant "The feeder on number 2 is hanging up a little this morning, so we'd better look at it." This person should be encouraged to take a walk around the factory at the beginning of every shift to sense what is going on and inspect any questionable events. The human senses of an experienced person are the best detection systems available today.

How To Start

The necessary items for establishing an effective preventive maintenance program are:

1. Every equipment uniquely identified by prominent ID number or serial number and product type
2. Accurate equipment history records
3. Failure information by problem/cause/action
4. Experience data from similar equipment
5. Manufacturer's interval and procedure recommendations
6. Service manuals
7. Consumables and replaceable parts
8. Skilled personnel
9. Proper test instruments and tools
10. Clear instructions with a checklist to be signed off
11. User cooperation
12. Management support.

A typical initial challenge is to get proper documentation for all equipment. When a new building or plant is constructed, the architects and construction engineers should be required to provide complete documentation on all facilities and the equipment installed in them. Any major equipment that is installed after that should have complete documentation. Figure 3.4 is a checklist that should be given to anyone who purchases facilities and equipment that must be maintained. As can be seen, one of the items on this list is ensuring availability of complete documentation and preventive maintenance recommendations.

Purchasing agents and facilities engineers are usually pleased to have such a checklist and will be cooperative if reminded occasionally about their major influence on life-cycle costs. This brings us back again to the principle of avoiding or minimizing the need for maintenance. Buying the right equipment

	Yes	No	Comments
1. Standardization			
a. Is equipment already in use that provides the desired function?			
b. Is this the same as existing equipment?			
c. Are there problems with existing equipment?			
d. Can we maintain this equipment with existing personnel?			
e. Are maintenance requirements compatible with our current procedures?			
2. Reliability and Maintainability			
a. Can vendor prove the equipment will operate at least to our specifications?			
b. Warranty of all parts and labor for 90+ days?			
c. Is design fault-tolerant?			
d. Are tests go/no go?			
3. Service Parts			
a. Is recommended replacement list provided?			
b. Is the dollar total of spares less than 10% of equipment cost?			
c. Do we already have usable parts?			
d. Can parts be purchased from other vendors?			
e. Are any especially high quality or expensive parts required?			
4. Training			
a. Is special technician training required?			
b. Will manufacturer provide training?			
1. At no additional cost for first year?			
2. At our location as required?			

Figure 3.4 Maintenance considerations checklist for purchasing agents and facilities engineers.

(continues)

	Yes	No	Comments
5. Documentation			
a. All technical manuals provided?			
1. Installation			
2. Operation			
3. Corrective and preventive maintenance			
4. Parts			
6. Special Tools and Test Equipment			
a. Do we already have all required tools and test equipment?			
b. Can at least 95% of all faults be detected by use of proposed equipment?			
c. Are calibration procedures minimum and clear?			
7. Safety			
a. Are all UL/SCA, OSHA, EPA and other applicable requirements met?			
b. Are any special precautions required?			
c. Can one person do all maintenance?			

Figure 3.4 cont'd.

in the beginning is the way to start. The best maintainability is eliminating the need for maintenance.

If you are in the captive service business or concerned with designing equipment that can be well maintained, you should recognize that the preceding has been aimed more at factory maintenance; but after all, that is an environment in which your equipment will often be used. It helps to view the program from the operator and service person's eyes to ensure that everyone's needs are satisfied.

4

PLANNING AND SCHEDULING

Planning is the heart of good inspection and preventive maintenance. As described earlier, the first thing to establish is what items must be maintained and what the best procedure is for performing that task. Establishing good procedures requires a good deal of time and talent. This can be a good activity for a new graduate engineer, perhaps as part of a training process that rotates him or her through various disciplines in a plant or field organization. This experience can be excellent training for a future design engineer.

Writing ability is an important qualification, along with pragmatic experience in maintenance practices. The language used should be clear and concise, with short sentences. Who, what, when, where, why, and how should be clearly described. A typical preventive maintenance procedure is illustrated in Figure 4.1. The following points should be noted from this typical procedure:

1. Every procedure has an identifying number and title.
2. The purpose is outlined.
3. Tools, reference documents, and any parts are listed.
4. Safety and operating cautions are prominently displayed.
5. A location is clearly provided for the maintenance mechanic to indicate performance as either satisfactory or deficient. If it is deficient, details are written in the space provided at the bottom for planning further work.

The procedure may be printed on a reusable, plastic-covered card that can be pulled from the file, marked, and returned when the work order is complete; on a standard preprinted form; or on a form that is uniquely printed by computer each time a related work order is prepared.

Truck 3500 Mile Oil Change

PURPOSE:	List cautions and steps required for changing oil.
REFERENCE:	Driver's manual for vehicle.
	Ensure vehicle is blocked securely before going under it.
CAUTIONS:	Hot oil from a recently operating motor can burn.
	Ensure adequate ventilation when running gas or diesel engine.
PROCEDURES:	
	Get replacement oil from stockroom.
	Get tools: catch basin, oil spout, wrench, wipes.
	Run motor at least 3 minutes to warm oil and mix contaminant particles.
	Position vehicle on grease rack, lift, or oil change station.
	Assure lift lock, blocks, and all safety devices are in safe position.
	Position catch basin under oil drain.
	Remove drain plug with wrench and drain oil into catch basin.
	When oil slows to a trickle, replace drain plug.
	If engine has a second sump, drain it the same way.
	Open hood, remove oil fill cap, and fill engine with fresh oil.
	Run engine 1 minute to circulate oil. Check underneath for any leaks.
	Check dipstick to assure oil level indicates in full area.
	Clean any spilled oil.
	Close hood and clean off any oil or fingerprints.
	Remove any old stickers from driver's door hinge column.
	Fill out oil change sticker with mileage and stick inside driver's door hinge column.
	Drive vehicle to parking area. Be alert for indications of other problems.
	Sign and date this checklist and write in mileage.

Completed by:		Date
Vehicle I.D.#:	License:	Odometer miles:

Figure 4.1 A typical preventive maintenance procedure.

Whatever the medium of the form, it should be given to the preventive maintenance craftsperson together with the work order so that he has all the necessary information at his fingertips. The computer version has the advantage of single-point control that may be uniformly distributed to many locations. This makes it easy for an engineer at headquarters to prepare a new procedure or to make any changes directly on the computer and have them instantly available to any user in the latest version.

Two slightly different philosophies exist for accomplishing the unscheduled actions that are necessary to repair defects found during inspection and preventive maintenance. One is to fix them on the spot. The other is to identify them clearly for later corrective action. This logic is outlined in Figure 4.2. If a "priority one" defect that could hurt a person or cause severe damage is

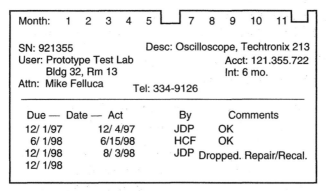

Figure 4.2 Logic for inspection findings.

observed, the equipment should be immediately stopped and "red tagged" so that it will not be used until repairs are made. Maintenance management should establish a guideline such as, "Fix anything that can be corrected within 10 minutes, but if it will take longer, write a separate work request." The policy time limit should be set, based on

1. Travel time to that work location
2. Effect on production
3. Need to keep the craftsperson on a precise time schedule.

The inspector who finds them can affect many small repairs most quickly. This avoids the need for someone else to travel to that location, identify the problem, and correct it. And it provides immediate customer satisfaction. More time-consuming repairs would disrupt the inspector's plans, which could cause other, even more serious problems to go undetected. The inspector is like a general practitioner who performs a physical exam and may give advice on proper diet and exercise but who refers any problems he may find to a specialist.

The inspection or preventive maintenance procedure form should have space where any additional action required can be indicated. When the procedure is completed and turned in to maintenance control, the planner or scheduler should note any additional work required and see that it gets done according to priority.

ESTIMATING TIME

Since inspection or preventive maintenance is a standardized procedure with little variation, the tasks and time required can be accurately estimated. Methods of developing time estimates include consideration of such resources as the following:

1. Equipment manufacturers' recommendations
2. National standards such as Chilton's on automotive or Means' for facilities
3. Industrial engineering time-and-motion studies
4. Historical experience.

Experience is the best teacher, but it must be carefully critiqued to make sure that the "one best way" is being used and that the pace of work is reasonable.

The challenge in estimating is to plan a large percentage of the work (preferably at least 90%) so that the time constraints are challenging but achievable without a compromise in high quality. The tradeoff between reasonable time and quality requires continuous surveillance by experienced supervisors. Naturally, if a maintenance mechanic knows that his work is being time studied, he will follow every procedure specifically and will methodically check off each step of the procedure. When the industrial engineer goes away, the mechanic will do what he feels are necessary items in an order that may or may not be satisfactory. As has been discussed in earlier, regarding motivation, an experienced preventive maintenance inspector mechanic can vary performance as much as 50% either way from standard, without most maintenance supervisors recognizing a problem or opportunity for improvement. Periodic checking against national or time-and-motion standards, as well as trend analysis of repetitive tasks, will help keep preventive task times at a high level of effectiveness.

ESTIMATING LABOR COST

Cost estimates follow from time estimates simply by multiplying the hours required by the required labor rates. Beware of coordination problems where multiple crafts are involved. For example, one "Fortune 100" company has trade jurisdictions that require the following personnel in order to remove an electric motor: a tinsmith to remove the cover, an electrician to disconnect the electrical supply, a millwright to unbolt the mounts, and one or more laborers to remove the motor from its mount. That situation is fraught with inefficiency and high labor costs, since all four trades must be scheduled together, with at least three people watching while the fourth is at work. The cost will be at least four times what it could be and is often greater if one of the trades does not show up on time. The best a scheduler can hope for is, if he has the latitude, to schedule the cover removal at say, 8:00 a.m., and the other functions at reasonable time intervals thereafter: electrician at 9:00, millwright at 10:00, and laborers at 11:00.

It is recommended that estimates be prepared on "pure" time. In other words, the exact hours and minutes that would be required under perfect scheduling conditions should be used. Likewise, it should be assumed that equipment would be available from production immediately. Delay time should be reported and scheduling problems should be identified so that they can be addressed separately from the hands-on procedure times. Note that people think in hours and minutes, so 1 hour and 10 minutes is easier to understand than 1.17 hours.

ESTIMATING MATERIALS

Most parts and materials that are used for preventive maintenance are well known and can be identified in advance. The quantity of each item planned should be multiplied by the cost of the item in inventory. The sum of those extended costs will be the material cost estimate. Consumables such as transmission oil should be enumerated as direct costs, but grease and other supplies used from bulk should be included in overhead costs.

Feedback From Actual

The time and cost required for every work order should be reported and analyzed to provide guidance for more accurate planning in future. It is important to determine what causes the task and times to change. Blindly assuming that the future will be like the past, or even that the past was done perfectly, may be an error. Comparisons should certainly be made between different individuals doing the same tasks to evaluate results in the amount of time required, what was accomplished during that time, quality of workmanship, and equipment performance as a result of their efforts.

Some people will argue that setting time standards for preventive maintenance is counterproductive. They feel that the mechanic should be given as much time as he desires to ensure high-quality work. This is generally not true. In fact, the required tasks will generally expand or contract to fit the available time. Preventive maintenance inspection and lubrication can in fact be treated as a production operation with incentives for both time performance and equipment uptime capability. The standard maintenance estimating and scheduling techniques of time slotting, use of ranges, and calculations based on the log-normal distribution may be followed as reliable data and analytical competence are established. Since preventive maintenance time and costs will typically comprise 30–60% of the maintenance budget, accurate planning, estimating, and scheduling are crucial to holding costs and improving profits.

SCHEDULING

Scheduling is, of course, one of the advantages to doing preventive maintenance over waiting until equipment fails and then doing emergency repairs. Like many other activities, the watchword should be "PADA," which stands for "Plan-a-Day-Ahead." In fact, the planning for inspections and preventive activities can be done days, weeks, and even months in advance to assure that the most convenient time for production is chosen, that maintenance parts and materials are available, and that the maintenance workload is relatively uniform.

Scheduling is primarily concerned with balancing demand and supply. Demand comes from the equipment's need for preventive maintenance. Supply is the availability of the equipment, craftspeople, and materials that are necessary to do the work. Establishing the demand has been partially covered in the chapters on on-condition, condition monitoring, and fixed interval preventive maintenance tasks. Those techniques identify individual equipment as candidates for preventive maintenance.

Prioritizing

When the individual pieces of equipment have been identified for preventive maintenance, there must be a procedure for identifying the order in which they are to be done. Not everything can be done first. First In–First Out (FIFO) is one way of scheduling demand. Using FIFO means that the next preventive task picked off the work request list, or the next card pulled from the file, is the next preventive maintenance work order. The problem with this "first come, first served" method is that the more desirable work in friendly locations tends to get done while other equipment somehow never gets its preventive maintenance. The improved method is Priority = Need Urgency × Customer Rank × Equipment Criticality. The acronym NUCREC will help in remembering the crucial factors.

NUCREC improves on the Ranking Index for Maintenance Expenditures (RIME) in several ways:

1. The customer rank is added.
2. The most important item is given the number-one rating.
3. The number of ratings in the scale may be varied according to the needs of the particular organization.
4. Part essentiality may be considered.

A rating system of numbers 1 through 4 is recommended. Since most humans think of number 1 as the first priority to get done, the NUCREC system does number 1 first.

Need urgency ratings include

1. Emergency; safety hazard with potential further damage if not corrected immediately; call back for unsatisfactory prior work

2. Downtime; facility or equipment is not producing revenue
3. Routine and preventive maintenance
4. As convenient, cosmetic.

The customer ranks are usually as follows:

1. Top management
2. Production line with direct revenue implications
3. Middle management, research and development facilities, frequent customers
4. All others.

The equipment criticality ratings are as follows:

1. Utilities and safety systems with large area effect
2. Key equipment or facility with no backup
3. Most impact on morale and productivity
4. Low, little use or effect on output.

The product of the ratings gives the total priority. That number will range from 1 (which is $1 \times 1 \times 1$) to 64 ($4 \times 4 \times 4$). The lowest number work will be first priority. A "1" priority is a first-class emergency. When several work requests have the same priority, labor and materials availability, locations, and scheduling fit may guide which is to be done first.

The priorities should be set in a formal meeting of production and maintenance management at which the equipment criticality number is assigned to every piece of equipment. Similarly, a rank number should be applied to every customer and the need urgency should be agreed on. With these predetermined evaluations, it is easy to establish the priority for a work order either manually by taking the numbers from the equipment card and the customer list and multiplying them by the urgency or by having the computer do so automatically. Naturally, there may be a few situations in which the planner's judgment should override and establish a different number, usually a lower number so that the work gets done faster.

Ratings may rise with time. A good way to ensure that preventive maintenance gets done is to increase the need urgency every week. In a computer system that starts with preventive maintenance at 3, a preventive task that is to be done every month or less frequently can be elevated after one week to a 2, and finally to a 1 rating. Those increases should ensure that the preventive task is done within a reasonable amount of time. If preventive maintenance is required more often, the incrementing could be done more rapidly.

Dispatch of the preventive maintenance work orders should be based on the demand ordered by priority, consistent with availability of labor and materials. As discussed earlier, predictive maintenance provides a good buffer activity in service work, since time within a few days is not normally critical. The NUCREC priority system helps ensure that the most important items are done first.

Some pressure will be encountered from production people who want a particular work request filled right away instead of at the proper time in the priority sequence. It can be helpful to limit the "criticality 1" equipment and "rank 1" customers to 10%, since, according to Pareto's Principle of the Critical Few, they will probably account for the majority of activity. If rank 2 is the next 20%, rank 3 is 30%, and the balance is 40% for rank 4, the workload should be reasonably balanced. If temporary work needs exist for selected equipment or a customer needs to be given a higher priority, then equipment should be moved to a lower criticality for each equipment that is moved higher. After all, one objective of prioritization is to ensure that work gets done in proper sequence. A preventive maintenance action that is done on time should ensure that equipment keeps operating and that emergency work is not necessary.

Coordination with Production

Equipment is not always available for preventive maintenance just when the maintenance schedulers would like it to be. An overriding influence on coordination should be a cooperative attitude between production and maintenance. This is best achieved by a meeting between the maintenance manager and production management, including the foreman level, so that what will be done to prevent failures, how this will be accomplished, and what production should expect to gain in uptime may all be explained.

The cooperation of the individual machine operators is of prime importance. They are on the spot and most able to detect unusual events that may indicate equipment malfunctions. Once an attitude of general cooperation is established, coordination should be refined to monthly, weekly, daily, and possibly even hourly schedules. Major shutdowns and holidays should be carefully planned so any work that requires "cold" shutdown can be done during those periods. Maintenance will often find that they must do this kind of work on weekends and holidays, when other persons are on vacation. Normal maintenance should be coordinated according to the following considerations:

1. Maintenance should publish a list of all equipment that is needed for inspections, preventive maintenance, and modifications and the amount of cycle time that such equipment will be required from production.

2. A maintenance planner should negotiate the schedule with production planning so that a balanced workload is available each week.
3. By Wednesday of each week, the schedule for the following week should be negotiated and posted where it is available to all concerned; it should be broken down by days.
4. By the end of the day before the preventive activity is scheduled, the maintenance person who will do the preventive maintenance should have seen the first-line production supervisor in charge of the equipment to establish a specific time for the preventive task.
5. The craftsperson should make every effort to do the job according to schedule.
6. As soon as the work is complete, the maintenance person should notify the production supervisor so that the equipment may be put back into use.

Overdue work should be tracked and brought up to date. Preventive maintenance scheduling should make sure that the interval is maintained between preventive actions. For example, if a preventive task for May is done on the 30th of the month, the next monthly task should be done during the last week of June. It is foolish to do a preventive maintenance task on May 30 and another June 1, just to be able to say one was done each month. In the case of preventive maintenance, the important thing is not the score but how the game was played.

Opportunity Preventive Maintenance Activities

It is often helpful to do preventive maintenance when equipment suddenly becomes available, which may not be on a regular schedule. One method called Techniques of Routine Interim Maintenance (TRIM) was covered in the preceding section. TRIM means generally that specified cleaning, inspection, lubrication, and adjustments are done at every service call. TRIM can be very effective.

Another variation is to convert (or expand) a repair call to include preventive activities. A good work order or service call system will quickly show any preventive maintenance, modification, or other work due when equipment work is requested. The system should also check parts availability and print pick lists. Parts required for preventive maintenance replacement can then be taken to the site and all work done at one time. Unless production is in a hurry to use the equipment again as soon as possible, doing all work on a piece of equipment during the single access is much more efficient than having to gain access several times to perform a few tasks each time.

ENSURING COMPLETION

A formal record is desirable for every inspection and preventive maintenance job. If the work is at all detailed, a checklist should be used. The completed checklist should be returned to the maintenance office on completion of the work. Any open preventive maintenance work orders should be kept on report until the supervisor has checked the results for quality assurance and signed off approval. Modern computer technology with handheld computers and pen-based electronic assistants permit paperless checklists and verification. In many situations, a paper work order form is still the most practical medium for the field technician. The collected data should then be entered into a computer system for tracking.

5

SCHEDULED PREVENTIVE MAINTENANCE

When most people think of preventive maintenance, they visualize scheduled, fixed-interval maintenance that is done every month, every quarter, every season, or at some other predetermined intervals. That timing may be based on days, or on intervals such as miles, gallons, activations, or hours of use. The use of performance intervals is itself a step toward basing preventive tasks on actual need instead of just on a generality.

The two main elements of fixed-interval preventive maintenance are procedure and discipline. Procedure means that the correct tasks are done and the right lubricants applied and consumables replaced at the best interval. Discipline requires that all the tasks are planned and controlled so that everything is done when it should be done. Both of these areas deserve attention. The topic of procedures is covered in detail in the following sections.

Discipline is a major problem in many organizations. This is obvious when one considers the fact that many organizations do not have an established program. Further, organizations that do claim to have a program often fail to establish a good planning and control procedure to ensure accomplishment. Elements of such a procedure include the following:

1. Listing of all equipment and the intervals at which they must receive preventive maintenance
2. A master schedule for the year that breaks down tasks by month, week, and possibly even to the day

3. Assignment of responsible persons to do the work
4. Inspection by the responsible supervisor to make sure that quality work is done on time
5. Updating of records to show when the work was done and when the next preventive task is due
6. Follow-up as necessary to correct any discrepancies.

Note that there are variations within the general topic of scheduled fixed interval maintenance. Some tasks will be done every Monday whether or not they are necessary. Inspection may be done every Monday and preventive tasks done if a need is indicated. Seasonal maintenance may be directed by environmental changes rather than by strict calendar date. Use meters, such as an automobile odometer, allow quantitative measure of use that can be related to the parameters that will need to be maintained. One must consider the relationship of components to the meter readings; for example, a truck's need for maintenance will vary greatly depending on whether it is used for long hauls or for local deliveries. A truck that is started every few miles and driven in stop-and-go, dusty city conditions will need more frequent mileage maintenance than the same truck used for long trips of continuous driving.

Seasonal equipment such as air conditioners, lawn mowers, salt spreaders, and snow blowers require special maintenance care at the end of each season to clean and refurbish them and store them carefully so that they will not deteriorate and will be ready for the next season. A lawn mower, for example, should have all gasoline drained from the tank and then be run until it stops because it has completely run out of fuel. This ensures that gasoline is completely removed from the lines. Oil should be changed. The spark plug should be removed and cleaned. A tablespoon of engine oil should be poured into the cylinder through the spark plug hole and the cylinder pulled through several strokes to ensure that is well lubricated. The spark plug should be put back in its hole loosely. Grass, dirt, and other residue should be thoroughly cleaned from all parts of the mower. The blade should be sharpened and checked to see that is in good balance. The mower should be stored in a dry place until it is needed again. Then, when the grass starts growing, all one has to do is fill the tank, tighten the spark plug, and connect the ignition wire. The motor should start on the second try. Careful preparation of equipment for storage will pay a major dividend when the equipment is needed in a hurry.

LUBRICATION

Friction of two materials moving relative to each other causes heat and wear. Great Britain has calculated that friction-related problems cost their industries over $1 billion per annum. They coined a new term, *tribology* (derived from the

Greek word *tribos*, which means "rubbing"), to refer to new approaches to the old dilemma of friction, wear, and the need for lubrication.

Technology intended to improve wear resistance of metal, plastics, and other surfaces in motion has greatly improved over recent years, but planning, scheduling, and control of the lubricating program is often reminiscent of a plant handyman wandering around with his long-spouted oil can.

Anything that is introduced onto or between moving surfaces to reduce friction is called a *lubricant*. Oils and greases are the most commonly used substances, although many other materials may be suitable. Other liquids and even gases are being used as lubricants. Air bearings, for example, are used in gyroscopes and other sensitive devices in which friction must be minimal. The functions of a lubricant are to

1. Separate moving materials from each other to prevent wear, scoring, and seizure
2. Reduce heat
3. Keep out contaminants
4. Protect against corrosion
5. Wash away worn materials.

Good lubrication requires two conditions: sound technical design for lubrication and a management program to ensure that every item of equipment is properly lubricated.

Lubrication Program Development

Information for developing lubrication specifications can come from four main sources:

1. Equipment manufacturers
2. Lubricant vendors
3. Other equipment users
4. Individuals' own experience.

As with most other preventive maintenance elements, initial guidance on lubrication should come from manufacturers. They should have extensive experience with their own equipment both in their test laboratories and in customer locations. They should know which parts wear and are frequently replaced. Therein lies a caution—a manufacturer could in fact make short-term profits by selling large numbers of spare parts to replace worn ones. Over the long term, however, that strategy will backfire and other vendors, whose equipment is less prone to wear and failure, will replace them.

Lubricant suppliers can be a valuable source of information. Most major oil companies will invest considerable time and effort in evaluating their customers' equipment to select the best lubricants and frequency or intervals for change. Figure 5.1 shows a typical report. Naturally, the vendor hopes that the consumer

LUBRICATION CHART
PREPARED BY
BEALE OIL COMPANY

NAME ___SERVICE INFOSYSTEMS, INC.___   DATE ___2/10/99___

EQUIPMENT			LUBRICANTS RECOMMENDED
ELECTRICAL DEPARTMENT			
Electric Motors			
Bearings	Ring Oiled	Ck W/Ch Y	MOBIL D.T.E. Oil Heavy Medium
Bearings	Hand Oiled	M	MOBIL D.T.E. Oil Heavy Medium
Bearings	Greased	2/Y	MOBILPLEX EP No. 1
Couplings	Greased	2/Y	MOBILPLEX EP No. 1
Motor—Generator Sets			
Bearings	Ring Oiled	Ck W/Ch Y	MOBIL D.T.E. Oil Heavy Medium
Bearings	Hand Oiled	M	MOBIL D.T.E. Oil Heavy Medium
Bearings	Greased	2/Y	MOBILPLEX EP No. 1
Couplings	Greased	2/Y	MOBILPLEX EP No. 1
York Refrigeration Compressor			
Brunner Refrigeration Compressor			
Crankcase & Cylinders	Splash		MOBIL D.T.E. Oil Heavy Medium
Transformers			
Circuit Breakers			
Compensators			
Insulating Oil			MOBILECT 25
Gear Motors			
Parallel Shafts	Splash	Ck M/Ch Y	MOBIL D.T.E. Oil Extra Heavy
Right Angle Worm Gears	Splash	Ck M/Ch Y	MOBIL 600 W Cylinder Oil
W — Weekly			
M — Monthly			
Y — Yearly			
2/Y — Twice Yearly			
Ck — Check			
Ch — Change			

Figure 5.1 Recommended lubricants.

will purchase his lubricants, but the total result can be beneficial to everyone. Lubricant vendors perform a valuable service of communicating and applying knowledge gained from many users to their customers' specific problems and opportunities.

Experience gained under similar operating conditions by other users or in your own facilities can be one of the best teachers. Personnel, including operators and mechanics, have a major impact on lubrication programs. Table 5.1 shows typical codes for methods of lubrication, intervals, actions, and responsibility. Figure 5.2 shows a typical lubrication schedule. Detailing of specific lubricants and intervals will not be done here, since they can be more effectively handled by the sources listed above.

The quality and the quantity of the lubricant applied are the two important conditions of any lube program. Lubrication properties must be carefully

Table 5.1 Lubrication Codes

Methods of Application

ALS	Automatic lube system
ALL	Air line lubricator
BO	Bottle oilers
DF	Drip feed
GC	Grease cups
GP	Grease packed
HA	Hand applied
HO	Hand oiling
ML	Mechanical lubricator
MO	Mist oiler
OB	Oil bath
OC	Oil circulation
OR	Oil reservoir
PG	Pressure gun
RO	Ring oiled
SLD	Sealed
SFC	Sight feed cups
SS	Splash system
WFC	Wick feed oil cups
WP	Waste packed

Servicing Actions

CHG	Change
CL	Clean
CK	Check
DR	Drain
INS	Inspect
LUB	Lubricate

Servicing Intervals

H	Hourly
D	Daily
W	Weekly
M	Monthly
Y	Yearly
NOP	When not operating
OP	OK to service when operating

Service Responsibility

MAE	Maintenance electricians
MAM	Maintenance mechanics
MAT	Maintenance trades
OPR	Operating personnel
OIL	Oiler

```
CARBON PLANT AREA     LUBRICATION  SCHEDULE  WEEK OF 12/15/99      AREA CODE   17
SCHEDULE LUBE DATE                                                 PAGE        23
12/28/99   SHIFT 2

SEQUENCE   LUBE POINT                        TYPE OF              OIL
 NUMBER    DESCRIPTION    METHOD   PLANTS   LUBRICATION  COMPLETED ADDED  CAPACITY

18-02   RACK & PINION     GUN       2       ANTI-SEIZE      Y N
18-03   SWIVEL            GUN       2       ANTI-SEIZE      Y N

************************** NO.20  ELEPHANT B 665-10 ***************************
20-01   BALL JOINT        GUN       2       ANTI-SEIZE      Y N
20-02   RACK & PINION     GUN       2       ANTI-SEIZE      Y N
20=03   SWIVEL            GUN       2       ANTI-SEIZE      Y N

************************** NO.21  ELEPHANT B-665-11 ***************************
22-01   BALL JOINT        GUN       2       ANTI-SEIZE      Y N
22-02   RACK & PINION     GUN       2       ANTI-SEIZE      Y N
22-03   SWIVEL            GUN       2       ANTI-SEIZE      Y N

                         - HOT PITCH STATION -

************************** PITCH PUMPS-3-INSIDE  ***************************
30-03   BEARINGS          GUN       4       ANTI-SEIZE      Y N

                         - 502 BUILDING -

************************** CONE CRUSHER-D-635-7  ***************************
38-00   CONE CRUSHER      RES       1        #5             Y N            35
38-02   BALL LOCK STOP    GUN       1        #1             Y N
38-03   BAND ROLL BRNGS   GUN       4        #1             Y N

                         - 503 BLDG GRND FLOOR MERZ -

************************** 640 FARVAL AUTO LUBE SYS ***********************
40-01   PUMP              DRUM            TEMP 78            Y N
40-02   AIRLINE OILER     RES             140               Y N

                         - 504 BUILDING GROUND FLOOR -

************************** KENNEDY VAN SAUN BALL MIL ***********************
44-01   DRUM END BRNGS    GUN       4       1               Y N
44-02   GEAR DRIVE &N&S   RES       2       10              Y N
```

Figure 5.2 Typical lubrication schedule.

selected to meet the operating conditions. The viscosity of the oil (or the base oil, if grease is used) and additives to provide film strength under pressure are especially important for bearing lubrication.

Too little lubricant is usually worse than too much, but excess can cause problems such as overheating and churning. The amount needed can range from a few drops per minute to a complete submersion bath.

A major step in developing the lubrication program is to assign specific responsibility and authority for the lubrication program to a competent maintainability or maintenance engineer. The primary functions and steps involved in developing the program are to:

1. Identify every piece of equipment that requires lubrication
2. Ensure that every major equipment is uniquely identified, preferably with a prominently displayed number

3. Ensure that equipment records are complete for manufacturer and physical location
4. Determine locations on each piece of equipment that needs to be lubricated
5. Identify lubricant to be used
6. Determine the best method of application
7. Establish the frequency or interval of lubrication
8. Determine if the equipment can be safely lubricated while operating or if it must be shut down
9. Decide who should be responsible for any human involvement
10. Standardize lubrication methods
11. Package the above elements into a lubrication program
12. Establish storage and handling procedures
13. Evaluate new lubricants to take advantage of state of the art
14. Analyze any failures involving lubrication and initiate necessary corrective actions.

Lubrication Program Implementation

An individual supervisor in the maintenance department should be assigned the responsibility for implementation and continued operation of the lubrication program. This person's primary functions are to

1. Establish lubrication service actions and schedules
2. Define the lubrication routes by building, area, and organization
3. Assign responsibilities to specific persons
4. Train lubricators
5. Ensure supplies of proper lubricants through the storeroom
6. Establish feedback that ensures completion of assigned lubrication and follows up on any discrepancies
7. Develop a manual or computerized lubrication scheduling and control system as part of the larger maintenance management program
8. Motivate lubrication personnel to check equipment for other problems and to create work requests where feasible
9. Ensure continued operation of the lubrication system.

It is important that a responsible person who recognizes the value of thorough lubrication be placed in charge. As with any activity, interest diminishes over time, equipment is modified without corresponding changes to the lubrication procedures, and state-of-the-art advances in lubricating technology may not be undertaken. A factory may have thousands of lubricating points that require attention. Lubrication is no less important to computer systems, even though they are often perceived as electronic. The computer field engineer must provide proper lubrication to printers, tape drives, and disks that spin at 3,600 rpm. A lot of maintenance time is invested in lubrication. The effect on production uptime can be measured nationally in billions of dollars.

CALIBRATION

Calibration is a special form of preventive maintenance whose objective is to keep measurement and control instruments within specified limits. A "standard" must be used to calibrate the equipment. Standards are derived from parameters established by the NBS. Secondary standards that have been manufactured to close tolerances and set against the primary standard are available through many test and calibration laboratories and often in industrial and university tool rooms and research labs. Ohmmeters are examples of equipment that should be calibrated at least once a year and before further use if subjected to sudden shock or stress.

Standards

The purpose of a calibration system is to provide for the prevention of tool inaccuracy through prompt detection of deficiencies and timely application of corrective action. Every organization should prepare a written description of its calibration system. This description should cover the measuring of test equipment and standards, including the following:

1. Establishment of realistic calibration intervals
2. Listing of all measurement standards
3. Establishment of environmental conditions for calibration
4. Ensuring the use of calibration procedures for all equipment and standards
5. Coordinating the calibration system with all users
6. Ensuring that equipment is frequently checked by periodic system or cross-checks to detect damage, inoperative instruments, erratic readings, and other performance-degrading factors that cannot be anticipated or provided for by calibration intervals
7. Providing for timely and positive correction action
8. Establishing decals, reject tags, and records for calibration labeling
9. Maintaining formal records to ensure proper controls.

Inspection Intervals

The checking interval may be in terms of time (hourly, weekly, monthly) or based on amount of use (every 5,000 parts, or every lot). For electrical test equipment, the power-on time may be critical factor and can be measured through an electrical elapsed-time indicator.

Adherence to the checking schedule makes or breaks the system. The interval should be based on stability, purpose, and degree of usage. If initial records indicate that the equipment remains within the required accuracy for successive calibrations, then the intervals may be lengthened. On the other hand, if equipment requires frequent adjustment or repair, the intervals should be shortened. Any equipment that does not have specific calibration intervals should be (1) examined at least every 6 months, and (2) calibrated at intervals of no longer than 1 year. Adjustments or assignment of calibration intervals should be done in such a way that a minimum of 95% of equipment or standards of the same type is within tolerance when submitted for regularly scheduled recalibration. In other words, if more than 5% of a particular type of equipment is out of tolerance at the end of its interval, then the interval should be reduced until less than 5% is defective when checked.

Control Records

A record system should be kept on every instrument, including the following:

1. History of use
2. Accuracy
3. Present location
4. Calibration interval and when due
5. Calibration procedures and necessary controls
6. Actual values of latest calibration
7. History of maintenance and repairs.

Figure 5.3 shows a typical calibration label.

Test equipment and measurement standards should be labeled to indicate the date of last calibration, by whom it was calibrated, and when the next calibration is due. When the size of the equipment limits the application of labels, an identifying code should be applied to reflect the serviceability and due date for next calibration. This provides a visual indication of the calibration serviceability status. Both the headquarters calibration organization and the instrument user

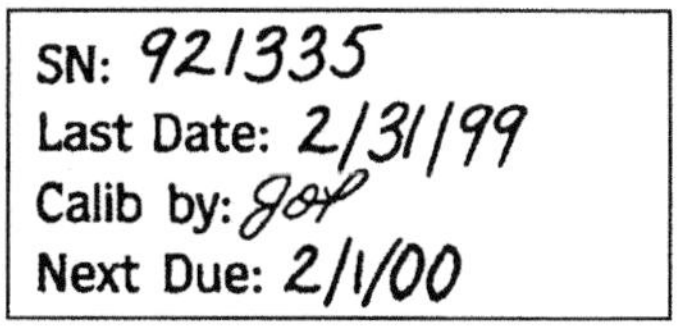

```
SN: 921335
Last Date: 2/31/99
Calib by: Joe
Next Due: 2/1/00
```

Figure 5.3 A typical calibration label.

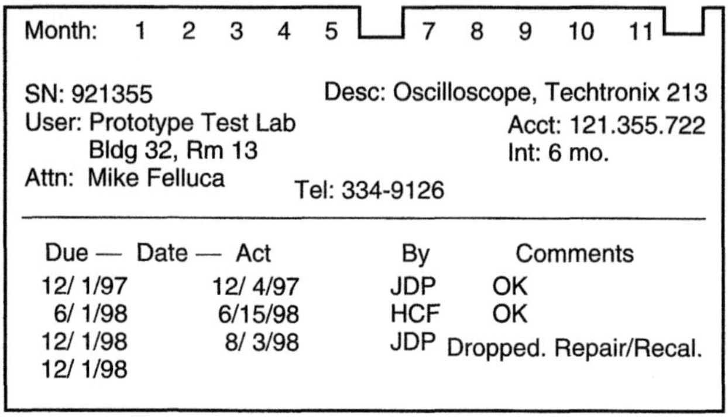

```
┌─────────────────────────────────────────────────────────┐
│ Month:   1   2   3   4   5 └─┘ 7   8   9   10   11 └─┘   │
│                                                          │
│ SN: 921355                Desc: Oscilloscope, Techtronix 213 │
│ User: Prototype Test Lab              Acct: 121.355.722  │
│        Bldg 32, Rm 13                 Int: 6 mo.         │
│ Attn: Mike Felluca       Tel: 334-9126                   │
│ ─────────────────────────────────────────────────────── │
│   Due — Date — Act         By        Comments           │
│   12/ 1/97      12/ 4/97    JDP     OK                   │
│    6/ 1/98       6/15/98    HCF     OK                   │
│   12/ 1/98       8/ 3/98    JDP  Dropped. Repair/Recal.  │
│   12/ 1/98                                               │
└─────────────────────────────────────────────────────────┘
```

Figure 5.4 A typical calibration card.

should maintain a two-way check on calibration. A simple means of doing this is to have a small form for each instrument with a calendar of weeks or months (depending on the interval required) across the top that can be punched and noted to indicate the calibration due date. An example of this sort of form is shown in Figure 5.4.

If the forms are sorted every month, the cards for each instrument that should be recalled for check or calibration can easily be pulled out.

6

MAINTENANCE ENGINEERING ROLES AND RESPONSIBILITIES

The primary reason for establishing a maintenance engineering function is to provide focus on asset reliability, maintainability, and life cycle cost for the entire facility. Therefore the roles, responsibilities, and accountability of this function must support these objectives. This fundamental requirement does not appear to be a part of the assigned roles and responsibilities for the Phillip Morris group. The observed deficiencies include:

1. The scope of equipment included in the maintenance engineering group's area of responsibility is limited to "production" equipment. By definition, or as interpreted, this excludes most of the infrastructure (electrical distribution, steam generation/distribution, compressed air/gases, etc.). In addition, other critical assets, such as cranes, are excluded.

2. Inconsistent vision of the true role of the maintenance engineering function. In its current configuration, there are 34 tasks or roles defined for the maintenance engineering function. While all of these appear to be valid activities, they do not fully define the role and responsibility of an effective functional group.

The role of an effective maintenance engineering function is to provide the proactive leadership, direction, and technical support required to achieve and sustain optimum reliability, maintainability, and life cycle cost for the facility's assets.

While maintenance engineering cannot directly affect facility performance, its responsibility is to provide facility and functional management with accurate, timely data that can be used to optimize maintenance and facility strategies that will support continuous improvement and ultimately result in world-class performance.

The responsibilities or duties that the function provides include:

❖ Develop criteria for effective maintenance management
- ➤ Methods to optimize maintenance strategy
  - Evaluate current practices versus best practices
  - Develop recommendations to correct deficiencies
  - Methods to validate preventive and corrective maintenance activities
  - Analyze preventive maintenance activities versus breakdown history by asset type, area, and classification
  - Evaluate and upgrade individual preventive maintenance task lists and work orders in suspect areas, asset types and classifications
- ➤ Methods to improve quality of work performed
  - Evaluate complete tasks, call-backs, reworks
  - Audit random executions of preventive and corrective activities
  - Skills assessments
- ➤ Methods to reduce maintenance workload
  - Evaluate maintenance history to determine proper periodicity and scope
  - Evaluate maintenance prevention methods that reduce maintenance requirements
  - Develop configuration management procedure to ensure reliability, maintainability, and best life cycle cost are followed in acquisition/modification of assets
  - Evaluate planning/scheduling effectiveness

❖ Improve asset reliability
- ➤ Ensure reliability and maintainability of new/modified installations
  - Develop configuration management procedure
  - Active participant in specification, procurement and installation of new assets or upgrades/modifications
  - Perform site acceptance tests, using predictive maintenance technologies, to verify inherent reliability of new/modified critical assets
  - Perform root-cause failure analysis on breakdowns and abnormal asset operation
- ➤ Identify and correct inherent design/installation/operation problems

- Perform simplified failure modes and effects analysis on critical assets
- Periodic evaluation of asset histories
- Periodic testing, using predictive technologies, to identify incipient reliability problems
- Verify and validate standard procedures (SOPs, PMs, and work orders)
- Develop/modify PMs and work orders for critical assets

➤ Improve life cycle costs

- Maintain and analyze equipment data and history records to predict future maintenance needs
- Develop effective procedures for inspection, adjustments, MRO parts, asset replacements, overhauls, etc. for critical assets
- Ensure assets are properly designed, selected, installed, operated, and maintained based on life cycle cost philosophy
- Monitor and evaluate asset performance
- Review asset deficiencies and implement corrections
- Perform periodic cost-benefit evaluations
- Identify and correct chronic and/or costly asset problems

➤ Provide technical support

- Maintenance manager
- Planners/schedulers
- Supervisors/foremen
- Maintenance crafts
- Procurement

➤ Plant Engineering

CONFIGURATION MANAGEMENT

Statistically, at least 85% of all reliability, asset utilization, and high life cycle cost problems are directly attributable to deficiencies in or total lack of enforced configuration management. Our database, as well as those developed by other consulting firms, indicates that the functional responsibility for these problems break out as follows:

- 23% caused by deficiencies in the production or operations function. The majority of these deficiencies are caused by a lack of valid, enforced operating procedures, poor skills, and unknown operating requirements of facility assets. A viable configuration management process could eliminate almost all of these forcing functions.
- 17% caused by deficiencies in the maintenance function. Again, the majority of these deficiencies are caused by a lack of effective or

enforced configuration management that predetermines the maintenance activities required to achieve and sustain asset reliability and that support best life cycle costs.

- 12% caused by deficiencies in the procurement process. In addition to mistakes made during the procurement of new capital assets, these deficiencies are caused by the lack of an effective procedure that governs the replacement of operating and maintenance spare parts. Again, an effective, enforced configuration management process would eliminate most of these deficiencies.

- 22% caused by plant or maintenance engineering. Most of these problems are caused by a lack of a formal procedure that can be used to evaluate the impact on reliability, availability, maintainability, and life cycle cost caused by modification or upgrades to existing facility assets. In addition, the lack of formal procedures directly and negatively affects the procurement of new or replacement assets.

- 11% caused by management philosophy. The majority of these deficiencies are caused by business decisions that are based on faulty data. Too many business decisions are made on opinions, perceptions, or intuitive judgments, and in most cases they are the wrong decisions. Implementing and enforcing an effective configuration management process could resolve at least one half of these deficiencies. The discipline and absolute adherence to standard procedures used to develop business plans, requests for capital expenditures, key performance indicators, and the myriad other information that senior managers rely on to make business decision will greatly increase the probability that the correct decision will be made.

- 15% caused by deficiencies in the sales and marketing function. The primary forcing function caused by sales is the method used to load the facility. The loading directly affects equipment utilization, production schedules, and maintenance activities. In effect, the way that sales loads the facility to a large degree determines the resultant reliability and life cycle cost of its assets. Normally, configuration management does not directly address the sales function's contribution to facility performance. However, procedures can and should be included that will minimize any negative effect that facility loading would cause.

Definitions of Configuration Management

There are two classic definitions of configuration management. The first is the methodology of effectively managing the life cycle of the major asset, in this case the shipyard. This type of configuration management governs the development of strategic and tactical plans that will optimize the useful life of the facility and is based on traditional life cycle management concepts.

Total Facility Configuration Management

Key components, as shown in Figure 6.1, of this form of configuration management include the following.

Program Management This component includes the management plan; the definition of the critical elements that comprise the facility; and the definition of all interfaces, databases, and procedures that are needed to support a life-of-facility management program.

Design Requirements This component establishes the design requirements, system and process boundaries, specific asset or equipment lists, and engineering design basis that must be maintained for the facility. The procedure clearly defines how each step of the design and/or change process will be performed.

Document and Control This component identifies all of the documents, document storage requirements, document controls and tracking, and retrieval

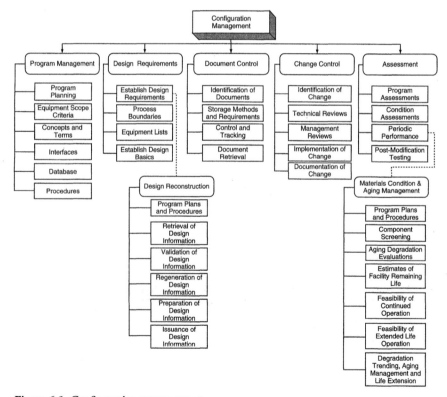

Figure 6.1 Configuration management.

requirements that are needed to support effective life cycle asset management. Documentation is a key requirement of effective configuration management. In a facility with effective configuration management, nothing can be done without proper documentation. Flying by the seat of your pants is simply not permitted.

Change Control This involves development and implementation of standard procedures to control configuration changes. The procedures provide specific methodology to identify, evaluate, manage, implement, and document changes.

Assessments The key to this type of configuration management is periodic assessments that quantify the condition of the shipyard and all of its assets. These assessments include physical configuration, criticality, condition, remaining useful life, life cycle costs, equipment performance (predictive maintenance), and other analyses or testing that quantify effectiveness.

Asset Condition and Aging Management This component is focused on useful life extension of the facility and its assets. It includes specific management methods and standard procedures that are designed to continuously evaluate asset condition and to develop effective means of extending useful life of assets. Generally, analyses that are used by this part of the configuration management process include aging degradation, feasibility of continued operation, and feasibility of extended operation.

This level of configuration management literally affects the entire organization. It provides standard procedures that define all aspects of day-to-day operations as well as the tactical and strategic planning process that will govern future actions. All of these procedures are predicated on the optimization of the shipyard for as long as it is feasible to continue operations.

Engineering Change Management

The second definition of configuration management, as illustrated in Figures 6.2 Level 1 and Level 2, is a subset of the first and is focused on effective management of the individual components (assets) that make up the shipyard. Normally, this process is known as engineering change management or life cycle costs management and governs all aspects of facility operations as they relate to the assets including all materials such as consumables, maintenance materials, drawings, training, etc. that directly or indirectly affect these assets. At the global level, these concepts are similar, but there are differences in the more detailed levels of the process. Both processes are intended to bring logic and discipline to the process of managing the life cycle cost of facility assets. An effective process should ensure that *all* decisions that directly or indirectly affect reliability,

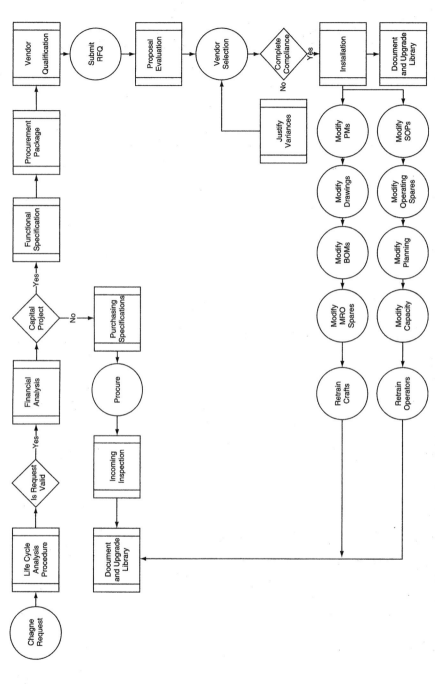

Figure 6.2 Level 1 Engineering change management.

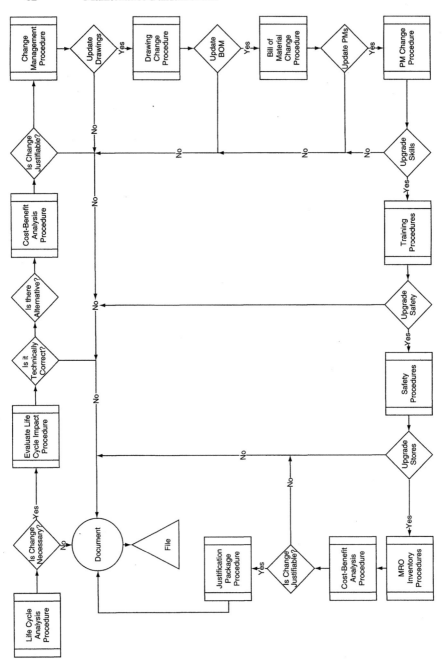

Figure 6.2 Level 2 Engineering change management.

maintainability, life cycle cost, and financial performance of the facility are based on best practices (i.e., thorough analysis based on factual data and a disciplined decision-making process).

Configuration Management for New Assets

Configuration management for the acquisition of new assets or major modification of existing assets must include specific procedures that define how to perform the following tasks:

1. Justify the need (for new or replacement systems, assets or equipment): All asset owners, engineering, and other function groups or individuals that are authorized to prepare a justification package will use this procedure. By using a standard procedure, senior management will be able to evaluate the real need for the recommended acquisitions.

2. Comprehensive engineering evaluation: A standard procedure that governs every step of the engineering evaluation for requests that are initially authorized by senior management. The procedure should include (1) technical analysis of requested system, asset or equipment; (2) evaluation of changes within the facility that will result from the change; and (3) development of a project plan to procure, implement and turn over the new system, asset or equipment, including all changes in documentation, training, procedures, capital spares, maintenance spares, etc. The procedure should also include the cost-benefit methodology that is needed to verify the need and the expected life cycle cost of the new asset.

3. Functional specifications: A procedure that governs the development of a comprehensive functional specification for the requested asset. This specification should include all of the data included in Procedure 2. This specification will be used for the procurement of the asset. It should include all labor and material requirements that should be provided by the vendor, contractors, and company. In addition, it should include all internal changes (i.e., training, drawings, procedures, spare parts inventory, etc.) caused by the inclusion of the new or replacement asset. The functional specification should include specific testing, acceptance, and documentation requirements that the vendor or others are to provide as part of the procurement or because of the procurement.

4. Procurement package: A procedure that ensures inclusion of all technical, financial, training, and other requirements that the vendor is expected to provide as part of the procurement. In addition, the package should include specific acceptance criteria, delivery dates,

penalties for off-specification or timeline, and other pertinent requirements that the vendor is expected to meet.

5. Qualified vendor selection: A specific procedure that is universally used to prequalify vendors for the procurement of assets, spares, consumables, and other materials and services that are needed to support the operation and maintenance of the facility. An effective prequalification procedure improves the potential for acquiring new or replacement assets that will support best life cycle cost.

6. Proposal evaluation: A formal procedure that governs the evaluation process for new and replacement assets. The procedure should include all steps required to evaluate the technical merit, life cycle cost, and long-term support that each potential vendor will provide as part of its proposal.

7. Proposal acceptance: The procedure that determines the logic that will be used to select the asset and vendor that will be procured. The acceptance criteria should be based on reliability, maintainability, and best projected life cycle cost.

8. Installation: A procedure that governs the methods and acceptance criteria that will be followed for the installation, testing, startup, and turnover of new and replacement assets. The procedure should include all probable combinations of turnkey, contractor, and in-house installations.

9. Acceptance: The engineering analysis, procurement package, and vendor submittal should include a specific method for factory and site acceptance testing that must be completed before the procured assets or system will be accepted.

10. Standard operating procedures: This procedure should include the methods that must be followed to ensure that all changes within the facility caused by the procurement of the new or replacement asset are made. Typically, these changes include modification or replacement of standard operating procedures, additional training, as well as possible changes in production materials.

11. Standard maintenance procedure: This procedure should include the methods that must be followed to ensure that all changes within the facility caused by the procurement of the new or replacement asset are made. As in operations, this includes standard procedures, preventive maintenance tasks, maintenance spares, training, drawings, bill of materials, special tools, and a variety of other changes.

12. Configuration change: This procedure defines the methodology that must be followed to make any change to installed assets or facility configuration. The procedure should include all modifications (i.e.,

bill of materials, drawings, operating/maintenance procedures, manning levels, required skills, etc.) that will result from the change. In addition, the procedure should provide clear procedures that require a thorough engineering and financial evaluation of any proposed change before it is submitted for approval as well as procedures for the approval and implementation process.

13. Decommissioning: This procedure governs the final decommissioning of the asset or assets when they reach the end of their useful life. The procedure should include removal, disposal, and possible replacement of the asset. It should provide specific instructions for assets or systems that may affect safety, environmental compliance, and other regulatory issues.

Configuration Management of Existing Assets

Configuration management for existing assets and governing direction for day-to-day operating of the facility should utilize most of the procedures defined above. Therefore the individual modules should be written to include specific directions for capital procurement, direct replacement of assets that do not justify a major capital procurement, changes in configuration (i.e., form, fit and function of facility systems), changes in mission, and other potential reasons that force a change in production or maintenance practices. In addition, the procedures should provide clear, concise procedures that are designed to prevent any change to the infrastructure or assets that is not based on thorough analysis of its effect the facility's ability to meet its mission requirement as well as reliability, maintainability, and life cycle cost.

The key to successful configuration management is documentation of all proposed and implemented changes and a universal (i.e., standard) methodology for implementing any change in the configuration or makeup of the facility. The level of documentation will be substantial and should be compatible with the SAP Enterprise information management system. As a point of information, SAP has a configuration management module that can be fully integrated into the facility's existing system. I have no direct experience with this module but understand that it will provide the means to ensure documentation control and facilitate the configuration management development process.

It should be noted that this description is only an overview of the myriad parameters that must be covered by an effective configuration management procedure. The level of detail required and the impact on most of the functional groups within the facility is substantial. However, any attempt to shortcut or simplify the process will seriously limit the benefits that could be derived.

REASONS FOR STANDARD PROCEDURES

The need for procedures is indicated when an organization is faced with decision options that may be precedent-setting. Situations calling for procedures include those in which

- Opinions may differ over the best course of action in situations that affect the achievement of overall mission and goals.
- Decisions may have significant consequences beyond the local level at which they are made.
- The choice of action may lead to unnecessary risk, counter-productivity, inefficiency, or conflict.
- Cooperation and reciprocal actions on the part of one organizational element are needed to enable another element to function effectively.

Procedures may be established to

Clarify	Define	Guide
Regulate	Direct	Establish
Integrate	Authorize	Enable
Empower	Commit	Support
Provide	Admit	Prohibit
Ensure	Inhibit	Standardize
Restrict	Disallow	

Reliability procedures are needed at a number of levels in the organization to ensure reliability creation, reliability maintenance, and reliability improvement. They are especially needed to smooth the way for key or critical matters. Firms establishing a new reliability function have a compelling need for them; firms engaged in the manufacture and sale of products or systems that can present a significant public safety hazard or can affect national security or national prestige have a critical need for them.

To be effective, reliability procedures must issue from high-level management. Management's attitude toward reliability, as expressed through procedures, is the most important single ingredient in making maintenance engineering and reliability assurance a successful practice in any organization. If the deeds of management in fact support written procedures, procedures gain credibility and legitimacy and will be respected.

Reliability procedures and, ultimately, the final responsibility for the reliability of products and services rest on the chief executive officer, albeit through successive management levels. The top assurance executive is responsible for pursuing reliability in the manner and to the extent prescribed by general management procedures and establishes his or her own procedures in support of this function.

Attributes of Procedures

Procedures, in general, should be as follows:

    a. Action-oriented (as contrasted to mere statement of belief)
    b. Supportive of organizational goals
    c. Consistent with other procedures
    d. Authoritative, credible, and acceptable at the level of implementation
    e. Inclusive to the extent of embracing all aspects of the intended application
    f. Specific to the extent of providing unambiguous direction and focus
    g. Admissive to the extent of allowing maximum flexibility of choice within the prescribed framework of guidance or direction
    h. Concise and readily understandable
    i. Relevant to the times and circumstances
    j. Stable over relatively long periods of time

Types and Levels of Procedures

Procedures are established at all levels of the organization, beginning at the corporate level and proceeding through divisional levels and successively finer levels of organizational structure down to departments and functional units. High-level procedures such as those at the corporate level are broad and general to deal effectively with the broad concerns of top-level management. Corporate reliability procedures provide for the establishment and promotion of reliability activity and achievement to fulfill obligations to customers and to society. They deal with internal matters pertaining to overall performance and with external matters pertaining to relationships with customers, the community, and involvements with regulatory and other such organizations. It sponsors the reliability function by declaring its intentions to the organization at large and provides for review and evaluation of the overall reliability system.

Divisional procedures respond to corporate procedures and relate to the more specific issues encountered by departments and functional units. Typically, they deal with administration, organizational interrelationships, operating methods, and the maintenance or improvement of organizational performance. Departmental procedures deal with situations and conditions more apt to arise on a day-to-day basis. These situations typically include matters relating to suppliers, subcontractors, product design activities, parts and materials, manufacturing, testing, auditing, and reviewing.

Developing and Establishing Procedures

The formation of effective and lasting procedures requires a comprehensive view of the issues and a full appreciation of the circumstances leading to the need for procedures. If the procedures developer is not in full possession of the facts and nuances, it is advisable to enlist the views of others who may be deeply involved in the situation and have the breadth of view, knowledge, judgment, and experience to make constructive contributions. It is desirable that managers and supervisors who would be affected by the procedures or who would be expected to carry them out should be considered as potential contributors. Not only can their views be constructive, their involvement in developing the procedures will increase their acceptance and support when the procedures go into effect. Although the participative approach is useful, the responsible manager nonetheless must impress the force of office and provide the benefit of experience to create appropriate procedures.

It is especially important to the formation of procedures for a new reliability function to win the acceptance and cooperation of long-established groups, particularly those that play prominent roles in the organization. Key individuals from those groups brought into the definition and development phases of procedures formation can help ensure well-conceived procedures by raising key issues to address.

While the development of many procedures is straightforward, some procedures are more involved and may require advance planning and study. To assist in such cases, the following outline is presented as a guide.

General Guide for Procedures Development

Effective procedures should follow well-defined guidelines that include the following:

1. State the need. Describe the situation that created the need. Identify who or what is involved, how they are involved, and to what extent.
2. Identify and review any existing procedures that relate to the situation.
3. Survey managers and supervisors who will be affected by the new procedures. Obtain pros and cons.
4. Determine if a new procedure is actually needed or if existing procedures should be revised to accommodate the situation.
5. Draft a preliminary procedure statement for review and comment by the departments affected. Include purpose and scope.

6. Integrate appropriate suggestions and prepare a revised statement for additional review. Add sections on responsibilities and actions, if appropriate.
7. Check the procedure against the attributes presented in the preceding section and prepare a final document.
8. With executive approval and sign off, release the document for distribution.

Procedures are communicated by memoranda, letters, instructions, or directives and are included in program plans and various manuals. They are given visibility by way of meetings, workshops, lectures, and training sessions. New procedures should be routed to all departments and units affected and acknowledged by signature and date.

Procedure statements range in size from simple statements of a paragraph or two in length to comprehensive documents that may include some or all of the following topics:

- Background
- Procedures
- Purpose
- Responsibility
- Scope
- Actions
- Definitions

While procedures are essential to effective and efficient operations, they should be held to the minimum allowed by size of organization; complexity of operations; criticality of processes; management style; and self-responsibility, awareness, and professional level of employees. As guiding forces, they should admit the widest possible latitude for action any given situation allows.

Procedures should be reviewed for possible revision or cancellation on a scheduled basis and whenever major changes are made in organization, management, practice, or overall organizational strategy.

The absence of a formal, universally followed configuration management procedure is a known cause of less-than-acceptable facility performance. Partial procurement specifications, as well as undocumented changes to installed assets, are a major contributor to chronic reliability problems that result in serious loss of capacity and excessive life cycle costs.

To the best of our knowledge, none of the required configuration management modules outlined above exist in the facility. Therefore the development process will literally be from a clean sheet of paper. Our best estimate is that will require a minimum of one (1) man-year to complete the initial procedure and will require the active participation of most of the functional groups that comprise the facility. As a minimum, the effort will require active input and/or participation from:

- Plant engineering
- Maintenance engineering
- Planning (business and maintenance)
- Procurement
- Labor relations
- Human resources
- Production
- Maintenance
- Customer relations
- Materials management
- Project management
- Asset owners
- Cost accounting.

Because the resultant procedure must include multiple functional groups (i.e., plant engineering, procurement, training, human resources, etc.), this task cannot be completed by maintenance engineering, the maintenance organization, or any other single functional group within the facility. The best way to proceed is to assign, with the approval of the functional managers, a cross-functional team to develop the procedure. As a minimum, this team should include representatives of plant engineering, maintenance engineering, project management, production, procurement, and financial management. Since these functional groups play the greatest role in the configuration management process, they are best suited to develop the procedures. The assigned team will need input from other individuals and/or groups but should be able to develop a mutually agreeable and effective procedure.

7

SHAFT ALIGNMENT

Shaft alignment is the proper positioning of the shaft centerlines of the driver and driven components (i.e., pumps, gearboxes, etc.) that make up the machine drive train. Alignment is accomplished through either shimming or moving a machine component or both. Its objective is to obtain a common axis of rotation at operating equilibrium for two coupled shafts or a train of coupled shafts.

Shafts must be aligned as perfectly as possible to maximize equipment reliability and life, particularly for high-speed equipment. Alignment is important for directly coupled shafts as well as coupled shafts of machines that are separated by distance—even those using flexible couplings. It is important because misalignment can introduce a high level of vibration, cause bearings to run hot, and result in the need for frequent repairs. Proper alignment reduces power consumption and noise level and helps to achieve the design life of bearings, seals, and couplings.

Alignment procedures are based on the assumption that one machine-train component is stationary, level, and properly supported by its base plate and foundation. Both angular and offset alignment must be performed in the vertical and horizontal planes, which is accomplished by raising or lowering the other machine components and/or moving them horizontally to align with the rotational centerline of the stationary shaft. The movable components are designated as machines to be moved (MTBM) or machines to be shimmed (MTBS). MTBM generally refers to corrections in the horizontal plane, while *MTBS* generally refers to corrections in the vertical plane.

Too often, alignment operations are performed randomly, and adjustments are made by trial and error, resulting in a time-consuming procedure. Because of this problem, Integrated Systems, Inc. (ISI) developed this module to help maintenance technicians speed up the alignment process. It presents a step-by-step procedure for the proper alignment of machinery and discusses shaft alignment fundamentals, equipment, measurements, and computations. Because there are certain basic math skills needed to perform alignment computations, a math review also is included.

FUNDAMENTALS

This section discusses the basic fundamentals of machine alignment and presents an alternative to the commonly used method, trial and error. This section addresses exactly what alignment is and the tools needed to perform it, why it is needed, how often it should be performed, what is considered to be "good enough," and what steps should be taken prior to performing the alignment procedure. It also discusses types of alignment (or misalignment), alignment planes, and why alignment is performed on shafts as opposed to couplings.

Shafts are considered to be in alignment when they are colinear at the coupling point. The term *colinear* refers to the condition when the rotational centerlines of two mating shafts are parallel and intersect (i.e., join to form one line). When this is the case, the coupled shafts operate just like a solid shaft. Any deviation from the aligned or colinear condition, however, results in abnormal wear of machine-train components such as bearings and shaft seals.

Variations in machine-component configuration and thermal growth can cause mounting-foot elevations and the horizontal orientations of individual drive-train components to be in different planes. Nevertheless, they are considered to be properly aligned as long as their shafts are colinear at the coupling point.

Note that it is important for final drive-train alignment to compensate for actual operating conditions because machines often move after start-up. Such movement is generally the result of wear, thermal growth, dynamic loads, and support or structural shifts. These factors must be considered and compensated for during the alignment process.

Tools most commonly used for alignment procedures are dial indicators, adjustable parallels, taper gauges, feeler gauges, small-hole gauges, and outside micrometer calipers.

Why Perform Alignment and How Often?

Periodic alignment checks on all coupled machinery are considered to be one of the best "tools" in a preventive maintenance program. Such checks are important because the vibration effects of misalignment can seriously damage a piece of equipment. Misalignment of more than a few thousandths of an inch can cause vibration that significantly reduces equipment life.

Although the machinery may have been properly aligned during installation or during a previous check, misalignment may develop over a very short period of time. Potential causes include foundation movement or settling, accidentally bumping the machine with another piece of equipment, thermal expansion, distortion caused by connected piping, loosened hold-down nuts, expanded grout, rusting of shims, etc.

Indications of misalignment in rotating machinery are shaft wobbling, excessive vibration (in both radial and axial directions), excessive bearing temperature (even if adequate lubrication is present), noise, bearing wear pattern, and coupling wear.

Trial-And-Error Versus Calculation

Many alignments are done by the trial-and-error method. Although this method may eventually produce the correct answers, it is extremely time-consuming and, as a result, it is usually considered "good enough" before it really is. Rather than relying on "feel" as with trial-and-error, some simple trigonometric principles allow alignment to be done properly with the exact amount of correction needed, either measured or calculated, taking the guesswork out of the process. Such accurate measurements and calculations make it possible to align a piece of machinery on the first attempt.

What Is Good Enough?

This is a difficult question to answer, because there are vast differences in machinery strength, speed of rotation, type of coupling, etc. It also is important to understand that flexible couplings do not cure misalignment problems—a common myth in industry. Although they may somewhat dampen the effects, flexible couplings are not a total solution.

An easy (perhaps too easy) answer to the question of what is good enough is to align all machinery to comply exactly with the manufacturers' specifications. However, the question of which manufacturers' specifications to follow must be answered, as few manufacturers build entire assemblies. Therefore an alignment is not considered good enough until it is well within all manufacturers' tolerances and a vibration analysis of the machinery in operation shows the vibration effects caused by misalignment to be within the manufacturers' specifications or accepted industry standards. Note that manufacturers' alignment specifications may include intentional misalignment during "cold" alignment to compensate for thermal growth, gear lash, etc. during operation.

COUPLING ALIGNMENT VERSUS SHAFT ALIGNMENT

If all couplings were perfectly bored through their exact center and perfectly machined about their rim and face, it might be possible to align a piece of machinery simply by aligning the two coupling halves. However, coupling eccentricity often results in coupling misalignment. This does not mean, however, that dial indicators should not be placed on the coupling halves to obtain alignment measurements. It does mean that the two shafts should be rotated simultaneously when obtaining readings, which makes the couplings an extension of the shaft centerlines whose irregularities will not affect the readings.

Although alignment operations are performed on coupling surfaces because they are convenient to use, it is extremely important that these surfaces and the shaft "run true." If there is any runout (i.e., axial or radial looseness) of the shaft and/or the coupling, a proportionate error in alignment will result. Therefore, prior to making alignment measurements, the shaft and coupling should be checked and corrected for runout.

ALIGNMENT CONDITIONS

There are four alignment conditions: perfect alignment, offset or parallel misalignment, angular or face misalignment, and skewed or combination misalignment (i.e., both offset and angular).

PERFECT ALIGNMENT

Two perfectly aligned shafts are colinear and operate as a solid shaft when coupled. This condition is illustrated in Figure 7.1. However, it is extremely rare for two shafts to be perfectly aligned without an alignment procedure being

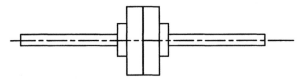

Figure 7.1 Perfect alignment.

performed on them. In addition, the state of alignment should be monitored on a regular basis to maintain the condition of perfect alignment.

OFFSET OR PARALLEL MISALIGNMENT

Offset misalignment, also referred to as *parallel misalignment*, refers to the distance between two shaft centerlines and is generally measured in thousandths of an inch. Offset can be present in either the vertical or horizontal plane. Figure 7.2 illustrates offset, showing two mating shafts that are parallel to each other but not colinear. Theoretically, offset is measured at the coupling centerline.

ANGULAR OR FACE MISALIGNMENT

A sound knowledge of angular alignment, also called *face misalignment*, is needed for understanding alignment conditions and performing the tasks associated with machine-train alignment, such as drawing alignment graphics, calculating foot corrections, specifying thermal growth, obtaining target specifications, and determining spacer-shaft alignment.

Angular misalignment refers to the condition when the shafts are not parallel but are in the same plane with no offset. This is illustrated in Figure 7.3. Note that with angular misalignment, it is possible for the mating shafts to be in the same plane at the coupling-face intersection but to have an angular relationship such that they are not colinear.

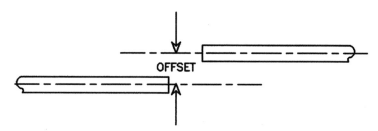

Figure 7.2 Offset misalignment.

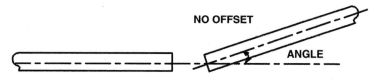

Figure 7.3 Angular misalignment (no offset).

Angularity is the angle between the two shaft centerlines, which generally is expressed as a "slope," or "rise over run," of so many thousandths of an inch per inch (i.e., unitless) rather than as an angle in degrees. It must be determined in both the vertical and horizontal planes. Figure 7.4 illustrates the angles involved in angular misalignment.

From a practical standpoint, it is often difficult or undesirable to position the stems of the dial indicators at 90-degree angles to the rim and/or face surfaces of the coupling halves. For this reason, brackets are used to mount the devices on the shaft or a non-movable part of the coupling to facilitate taking readings and to ensure greater accuracy. This is a valid method because any object that is securely attached and rotated with the shaft or coupling hub becomes a radial extension of the shaft centerline and can be considered an integral part of the shaft. However, this somewhat complicates the process and requires right-triangle concepts to be understood and other adjustments (e.g., indicator sag) to be made to the readings.

Compare the two diagrams in Figure 7.5. Figure 7.5a is a common right triangle and Figure 7.5b is a simplified view of an alignment-measuring apparatus, or fixture, that incorporates a right triangle.

The length of side "b" is measured with a tape measure and the length of side "a" is measured with a device such as a dial indicator. Note that this diagram assumes the coupling is centered on the shaft and that its centerline is the same as the shaft's. Angle "A" in degrees is calculated by

$$A = \tan^{-1}\frac{a}{b}$$

This formula yields the angle "A" expressed in degrees, which requires the use of a trigonometric table or a calculator that is capable of determining the inverse

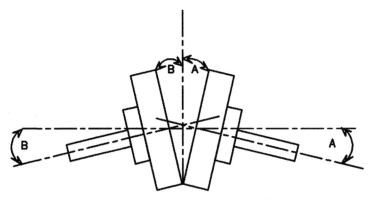

Figure 7.4 Angles are equal at the coupling or shaft centerline.

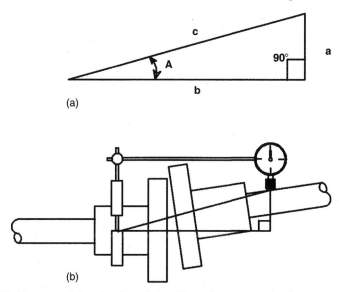

Figure 7.5 Common right triangle and simplified alignment-measuring apparatus.

tangent. Although technically correct, alignment calculations do not require the use of an angle value in degrees. Note that it is common industry practice to refer to the following value as "Angle-A," even though it is not truly an angle and is actually the tangent of Angle "A":

$$\text{"Angle-A"} = \frac{a}{b} = \frac{rise}{run}$$

Figures 7.6 and 7.7 illustrate the concept of rise and run. If one assumes that line O-A in Figure 7.6 represents a true, or target, shaft centerline, then side "a" of the triangle represents the amount of offset present in the actual shaft, which is referred to as the rise.

(Note that this "offset" value is not the true theoretical offset as defined in Chapter 2. It is actually the theoretical offset plus one-half of the shaft diameter (see Figure 7.5), because the indicator dial is mounted on the outside edge of the shaft as opposed to the centerline. However, for the purposes of alignment calculations, it is not necessary to use the theoretical offset or the theoretical run that corresponds to it. Figure 7.7 illustrates why this is not necessary.)

Figure 7.7 illustrates several rise/run measurements for a constant "Angle-A." Unless "Angle-A" changes, an increase in rise results in a proportionate increase in run. This relationship allows the alignment calculations to be made without using the theoretical offset value and its corresponding run.

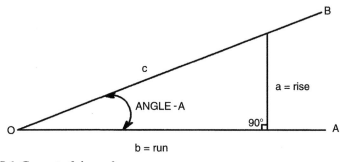

Figure 7.6 Concept of rise and run.

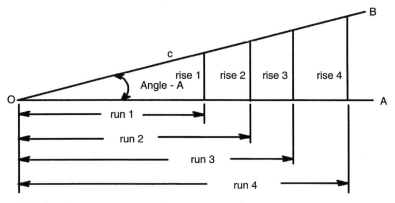

Figure 7.7 Rise/run measurements for constant angle.

Therefore, the calculation of "Angle-A" can be made with any of the rise/run measurements:

$$\text{"Angle-A"} = \frac{rise_1}{run_1} = \frac{rise_2}{run_2} = \frac{rise_3}{run_3} = \frac{rise_4}{run_4}$$

For example, if the rise at a machine foot is equal to 0.5 inches with a run of 12 inches, "Angle-A" is

$$\text{"Angle-A"} = \frac{0.5''}{12.0''} = 0.042$$

If the other machine foot is 12 inches away (i.e., run = 24 inches), the following relationship applies:

$$0.042 = \frac{X}{24.0''}$$

where X or rise = 1 inch

COMBINATION OR SKEWED MISALIGNMENT

Combination or skewed misalignment occurs when the shafts are not parallel (i.e., angular) nor do they intersect at the coupling (i.e., offset). Figure 7.8 shows two shafts that are skewed, which is the most common type of misalignment problem encountered. This type of misalignment can occur in either the horizontal or vertical plane, or in both the horizontal and vertical planes.

For comparison, see Figure 7.3, which shows two shafts that have angular misalignment but are not offset. Figure 7.9 shows how an offset measurement for non-parallel shafts can vary depending on where the distance between two shaft centerlines is measured. Again, note that theoretical offset is defined at the coupling face.

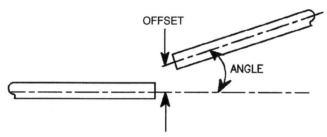

Figure 7.8 Offset and angular misalignment.

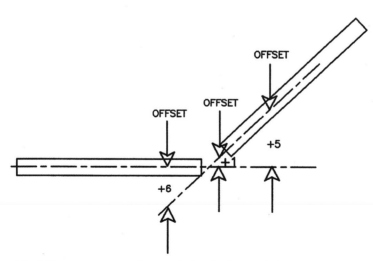

Figure 7.9 Offset measurement for angularly misaligned shafts.

ALIGNMENT PLANES

There are two misalignment planes to correct: vertical and horizontal. Therefore, in the case in which at least two machines make up a machine-train, four types of misalignment can occur: vertical offset, vertical angularity, horizontal offset, and horizontal angularity. These can occur in any combination, and in many cases, all four are present.

Vertical

Both angular misalignment and offset can occur in the vertical plane. Vertical misalignment, which is corrected by the use of shims, is usually illustrated in a side-view drawing as shown in Figure 7.10.

Horizontal

Both offset and angular misalignment can occur in the horizontal plane. Shims are not used to correct for horizontal misalignment, which is typically illustrated in a top-view drawing as shown in Figure 7.11. This type of misalignment is corrected by physically moving the MTBM.

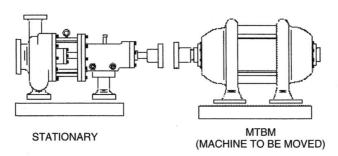

STATIONARY   MTBM
                    (MACHINE TO BE MOVED)

Figure 7.10 Vertical misalignment.

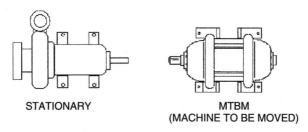

STATIONARY   MTBM
                    (MACHINE TO BE MOVED)

Figure 7.11 Horizontal misalignment.

ACTIONS TO BE TAKEN BEFORE ALIGNMENT

It is crucial that alignment procedures be performed correctly, regardless of what method from Chapter 3 is used. Actions to be taken before alignment are discussed in the following sections, which cover the preparatory steps as well as two major issues (i.e., soft-foot and indicator sag corrections) that must be resolved before alignment can be accomplished. This section provides procedures for making these corrections as well as the proper way to tighten hold-down nuts, an important procedure needed when correcting soft-foot.

Preparatory Steps

The following preparatory steps should be taken before attempting to align a machine train:

1. Before placing a machine on its base, make sure that both the base and the bottom of the machine are clean, rust free, and do not have any burrs. Use a wire brush or file on these areas if necessary.
2. Common practice is to position, level, and secure the driven unit at the required elevation prior to adjusting the driver to align with it. Set the driven unit's shaft centerline slightly higher than the driver.
3. Make all connections, such as pipe connections to a pump or output shaft connections on a reducer, to the driven unit.
4. Use only clean shims that have not been "kinked" or that do not have burrs.
5. Make sure the shaft does not have an indicated runout.
6. Before starting the alignment procedure, check for "soft-foot" and correct the condition.
7. Always use the correct tightening sequence procedure on the hold-down nuts.
8. Determine the amount of indicator sag before starting the alignment procedure
9. Always position the stem of the dial indicator so that it is perpendicular to the surface against which it will rest. Erroneous readings will result if the stem is not placed at a 90-degree angle to the surface.
10. Avoid lifting the machine more than is absolutely necessary to add or remove shims.
11. Jacking bolt assemblies should be welded onto the bases of all large machinery. If they are not provided, add them before starting the alignment procedure. Use jacking bolts to adjust for horizontal offset and angular misalignment and to hold the machine in place while shimming.

CORRECTING FOR SOFT-FOOT

Soft-foot is the condition when all four of a machine's feet do not support the weight of the machine. It is important to determine if this condition is present prior to performing shaft alignment on a piece of machinery. Not correcting soft-foot prior to alignment is a major cause of frustration and lost time during the aligning procedure.

The basis for understanding and correcting soft-foot and its causes is the knowledge that three points determine a plane. As an example, consider a chair with one short leg. The chair will never be stable unless the other three legs are shortened or the short leg is shimmed. In this example, the level floor is the "plane" and the bottom tips of the legs are the "points" of the plane. Three of the four chair tips will always rest on the floor. If a person is sitting with his or her weight positioned above the short leg, it will be on the floor and the normal leg diagonally opposite the short leg will be off the floor.

As in the chair example, when a machine with soft-foot is placed on its base, it will rest on three of its support feet unless the base and the bottoms of all of the feet are perfectly machined. Further, because the feet of the machine are actually square pads and not true points, it is possible that the machine can rest on two support feet, ones that are diagonally opposite each other. In this case, the machine has two soft-feet.

Causes

Possible sources of soft-foot are shown in Figure 7.12.

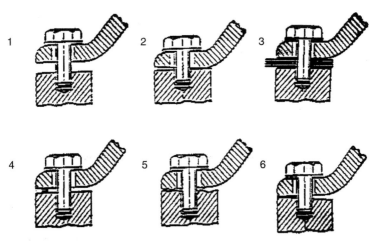

Figure 7.12 Diagrams of possible soft-foot causes. 1, Loose foot. 2, Cocked foot. 3, Bad shim. 4, Debris under foot. 5, Irregular base surface. 6, Cocked foot.

Consequences

Placing a piece of machinery in service with uncorrected soft-foot may result in the following:

- Dial-indicator readings taken as part of the alignment procedure can be different each time the hold-down nuts are tightened, loosened, and retightened. This can be extremely frustrating because each attempted correction can cause a soft-foot condition in another location.
- The nuts securing the feet to the base may loosen, resulting in either machine looseness and/or misalignment. Either of these conditions can cause vibration, which can be dangerous to personnel as well as to the machine.
- If the nuts do not loosen, metal fatigue may occur at the source of soft-foot. Cracks can develop in the support base/frame and, in extreme cases, the soft-foot may actually break off.

Initial Soft-foot Correction

The following steps should be taken to check for and correct soft-foot:

- Before setting the machine in place, remove all dirt, rust, and burrs from the bottom of the machine's feet, the shims to be used for leveling, and the base at the areas where the machine's feet will rest.
- Set the machine in place, but do not tighten the hold-down nuts.
- Attempt to pass a thin feeler gauge underneath each of the four feet. Any foot that is not solidly resting on the base is a soft-foot. (A foot is considered "soft" if the feeler gauge passes beneath most of it and only contacts a small point or one edge.)
- If the feeler gauge passes beneath a foot, install the necessary shims beneath that foot to make the "initial" soft-foot correction.

Final Soft-foot Correction

The following procedure describes the final soft-foot correction:

- Tighten all hold-down nuts on both the stationary machine and the MTBS.
- Secure a dial-indicator holder to the base of the stationary machine and the MTBS. The stem of the dial indicator should be in a vertical position above the foot to be checked. A magnetic-base indicator holder is most suitable for this purpose.
- Set the dial indicator to zero. Completely loosen the hold-down nut on the foot to be checked. Watch the dial indicator closely for foot movement during the loosening process.

- If the foot rises from the base when the hold-down nut is loosened, place beneath the foot an amount of shim stock equal to the amount of deflection shown on the dial indicator.
- Retighten the hold-down nut and repeat the entire process once again to ensure that no movement occurs.
- Move the dial indicator and holder to the next foot to be checked and repeat the process. Note: The nuts on all of the other feet must remain securely tightened when a foot is being checked for a soft-foot condition.
- Repeat the above process on all of the feet.
- Make a three-point check on each foot by placing a feeler gauge under each of the three exposed sides of the foot. This determines if the base of the foot is cocked.

Tightening Hold-Down Nuts

Once soft-foot is removed, it is important to use the correct tightening procedure for the hold-down nuts. This helps ensure that any unequal stresses that cause the machine to shift during the tightening procedure remain the same throughout the entire alignment process. The following procedure should be followed:

- After eliminating soft-foot, loosen all hold-down nuts.
- Number each machine foot in the sequence in which the hold-down nuts will be tightened during the alignment procedure. The numbers (1, 2, 3, and 4) should be permanently marked on, or near, the feet.
- It is generally considered a good idea to tighten the nuts in an "X" pattern as illustrated in Figure 7.13.
- Always tighten the nuts in the sequence in which the positions are numbered (1, 2, 3, and 4).
- Loosen nuts in the opposite sequences (4, 3, 2, and 1).
- Use a torque wrench to tighten all nuts with the same amount of torque.
- A similar procedure should be used for base plates.
- Always tighten the nuts as though the final adjustment has been made, even if the first set of readings has not been taken.

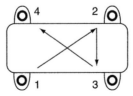

Figure 7.13 Correct bolting sequence for tightening nuts.

CORRECTING FOR INDICATOR SAG

Indicator sag is the term used to describe the bending of the mounting hardware as the dial indicator is rotated from the top position to the bottom position during the alignment procedure. Bending can cause significant errors in the indicator readings that are used to determine vertical misalignment, especially in rim-and-face readings (see Chapter 3). The degree to which the mounting hardware bends depends on the length and material strength of the hardware.

To ensure that correct readings are obtained with the alignment apparatus, it is necessary to determine the amount of indicator sag present in the equipment and to correct the bottom or 6 o'clock readings before starting the alignment process.

Dial indicator mounting hardware consists of a bracket clamped to the shaft, which supports a rod extending beyond the coupling. When two shafts are perfectly aligned, the mounting rod should be parallel to the axis of rotation of the shafts. However, the rod bends or sags by an amount usually measured in mils (thousandths of an inch) because of the combined weight of the rod and the dial indicator attached to the end of the rod. Figure 7.14 illustrates this problem.

Indicator sag is best determined by mounting the dial indicator on a piece of straight pipe of the same length as in the actual application. Zero the dial indicator at the 12 o'clock, or upright, position and then rotate 180 degrees to the 6 o'clock position. The reading obtained, which will be a negative number, is the measure of the mounting-bracket indicator sag for 180 degrees of rotation and is called the *sag factor*. All bottom or 6 o'clock readings should be corrected by subtracting the sag factor.

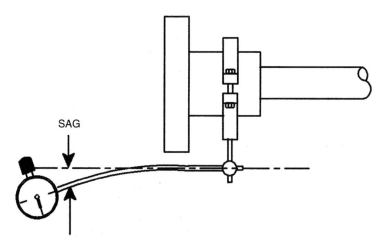

Figure 7.14 Dial indicator sag.

Example 1: Assume that the sag factor is −0.006 inch. If the indicator reading at 6 o'clock equals +0.010 inch, then the true reading is:

$$\text{Indicator reading} - \text{sag factor}$$
$$(+0.010'') - (-0.006'') = +0.016''$$

Example 2: If the indicator reading at 6 o'clock equals −0.010 inch, then the true reading is:

$$(-0.010'') - (-0.006'') = -0.004''$$

As shown by the above examples, the correct use of positive (+) and negative (−) signs is important in shaft alignment.

ALIGNMENT EQUIPMENT AND METHODS

There are two primary methods of aligning machine-trains: dial-indicator alignment and optical, or laser, alignment. This section provides an overview of each, with an emphasis on dial-indicator methods.

Dial-indicator methods (i.e., reverse dial indicator and the two variations of the rim-and-face method) use the same type of dial indicators and mounting equipment. However, the number of indicators and their orientations on the shaft are different. The optical technique does not use this device to make measurements but uses laser transmitters and sensors.

While the dial-indicator and optical methods differ in the equipment and/or equipment setup used to align machine components, the theory on which they are based is essentially identical. Each method measures the offset and angularity of the shafts of movable components in reference to a preselected stationary component. Each assumes that the stationary unit is properly installed and that good mounting, shimming, and bolting techniques are used on all machine components.

Dial-indicator Methods

There are three methods of aligning machinery with dial indicators. These methods are (1) the two-indicator method with readings taken at the stationary machine; (2) the two-indicator method with readings taken at the machine to be shimmed; and (3) the indicator reverse method. Methods 1 and 2 are often considered to be one method, which is referred to as *rim-and-face*.

Method Selection

Although some manufacturers insist on the use of the indicator reverse method for alignment or at least as a final check of the alignment, two basic factors determine which method should be used. The determining factors in method selection are (1) end play and (2) distance versus radius.

End Play or Float Practically all machines with journal or sleeve bearings have some end play or float. It is considered to be manageable if sufficient pressure can be applied to the end of the shaft during rotation to keep it firmly seated against the thrust bearing or plate. However, for large machinery or machinery that must be energized and "bumped" to obtain the desired rotation, application of pressure on the shaft is often difficult and/or dangerous. In these cases, float makes it impossible to obtain accurate face readings; therefore, the indicator reverse method must be used as float has a negligible effect.

Distance Versus Radius If float is manageable, then there is a choice of which of the methods to use. When there is a choice, the best method is determined by the following rule:

If the distance between the points of contact of the two dial indicators set up to take rim readings for the indicator reverse method is larger than one half the diameter of travel of the dial indicator set up to take face readings for the two-indicator method, the indicator reverse method should be used.

This rule is based on the fact that misalignment is more apparent (i.e., dial indicator reading will be larger) under these circumstances, and therefore corrections will be more accurate.

Equipment

Dial indicators and mounting hardware are the equipment needed to take alignment readings.

Dial Indicators Figure 7.15 shows a common dial indicator, which is also called a *runout gauge*. A dial indicator is an instrument with either jeweled or plain bearings, precisely finished gears, pinions, and other precision parts designed to produce accurate measurements. It is possible to take measurements ranging from one-thousandth (0.001 inch or one mil) to 50 millionths of an inch.

The point that contacts the shaft is attached to a spindle and rack. When it encounters an irregularity, it moves. This movement is transmitted to a pinion, through a series of gears, and on to a hand or pointer that sweeps the dial of the indicator. It yields measurements in (+) or (−) mils.

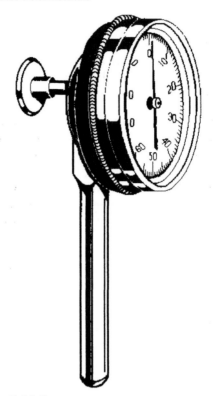

Figure 7.15 Common dial indicator.

Measurements taken with this device are based on a point of reference at the "zero position," which is defined as the alignment fixture at the top of the shaft—referred to as the 12 o'clock position. To perform the alignment procedure, readings also are required at the 3, 6, and 9 o'clock positions.

It is important to understand that the readings taken with this device are all relative, meaning they are dependent on the location at which they are taken. Rim readings are obtained as the shafts are rotated and the dial indicator stem contacts the shaft at a 90-degree angle. Face readings, which are used to determine angular misalignment, are obtained as the shafts are rotated and the stem is parallel to the shaft centerline and touching the face of the coupling.

Mounting Hardware Mounting hardware consists of the brackets, posts, connectors, and other hardware used to attach a dial indicator to a piece of machinery. Dial indicators can easily be attached to brackets and, because brackets are adjustable, they can easily be mounted on shafts or coupling hubs of varying size. Brackets eliminate the need to disassemble flexible couplings when checking alignments during predictive maintenance checks or when doing an actual alignment. This also allows more accurate "hot alignment" checks to be made.

The brackets are designed so that dial indicators can easily be mounted for taking rim readings on the movable machine and the fixed machine at the same time. This facilitates the use of the indicator reverse method of alignment. If there is not enough room on the shafts, it is permissible to attach brackets to the coupling hubs or any part of the coupling that is solidly attached to the shaft. Do not attach brackets to a movable part of the coupling, such as the shroud.

Note that misuse of equipment can result in costly mistakes. One example is the improper use of magnetic bases, which are generally designed for stationary service. They are not designed for direct attachment to a shaft or coupling that must be rotated to obtain the alignment readings. The shift in forces during rotation can cause movement of the magnetic base and erroneous readings.

Methods There are three primary methods of aligning machine-trains with dial indicators: reverse-dial indicator method, also called indicator-reverse method, and two variations of the rim-and-face method.

With all three of these methods, it is usually possible to attach two dial indicators to the machinery in such a manner that both sets of readings can be taken simultaneously. However, if only one indicator can be attached, it is acceptable to take one set of readings, change the mounting arrangement, and then take the other set of readings.

There are advantages with the reverse-dial indicator method over the rim-and-face method—namely, accuracy, and the fact that the mechanic is forced to perform the procedure "by the book" as opposed to being able to use "trial and error." Accuracy is much better because only rim readings are used. This is because rim readings are not affected by shaft float or end play as are face readings. In addition, the accuracy is improved as compared with rim-and-face methods because of the length of the span between indicators.

Reverse-dial Indicator Reverse-dial indicator method (also referred to as *indicator-reverse* method) is the most accurate form of mechanical alignment. This technique measures offset at two points, and the amount of horizontal and vertical correction for offset and angularity is calculated. Rim readings are taken simultaneously at each of the four positions (12, 3, 6, and 9 o'clock) for the movable machine (MTBS/MTBM) and the stationary machine. The measuring device for this type of alignment is a dual-dial indicator, and the most common configuration is that shown in Figure 7.16.

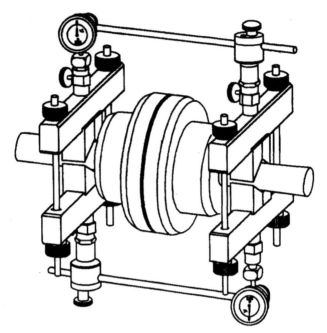

Figure 7.16 Typical reverse-dial indicator fixture and mounting.

Mounting Configuration and Readings

Dual runout gauges are rigidly mounted on special fixtures attached to the two mating shafts. The runout gauges are mounted so that readings can be obtained for both shafts with one 360-degree rotation.

When the reverse-dial fixture is mounted on mating shafts, the dials initially should be adjusted to their zero point. Once the dials are zeroed, slowly rotate the shafts in 90-degree increments. Record runout readings from both gauges, being sure to record the positive or negative sign, when the fixture is at the 12, 3, 6, and 9 o'clock positions.

Limitations

There are potential errors or problems that limit the accuracy of this alignment technique. The common ones include data recording errors, failure to correct for indicator sag, mechanical looseness in the fixture installation, and failure to properly zero and/or calibrate the dial indicator.

Data Recording One of the most common errors made with this technique is reversing the 3 and 9 o'clock readings. Technicians have a tendency to reverse their orientation to the machine-train during the alignment process. As a result, they often reverse the orientation of the recorded data.

To eliminate this problem, always acquire and record runout readings facing away from the stationary machine component. In this orientation, the 3 o'clock data are taken with the fixture oriented at 90 degrees (horizontal) to the right of the shafts. The 9 o'clock position is then horizontal to the left of the shafts.

Indicator Sag The reverse dial indicator fixture is composed of two mounting blocks, which are rigidly fixed to each of the mating shafts. The runout gauges, or dial indicators, are mounted on long, relatively small-diameter rods, which are held by the mounting blocks. As a result of this configuration, there is always some degree of sag or deflection in the fixtures. See Chapter 2 for a discussion on measuring and compensating for indicator sag.

Mechanical Looseness As with all measurement instrumentation, proper mounting techniques must be followed. Any looseness in the fixture mounting or at any point within the fixture will result in errors in the alignment readings.

Zeroing and Calibrating It is very important that the indicator dials be properly zeroed and calibrated before use. Zeroing is performed once the fixture is mounted on the equipment to be aligned at the 12 o'clock position. It is accomplished either by turning a knob located on the dial body or by rotating the dial face itself until the dial reads zero. Calibration is performed in the instrument lab by measuring known misalignments. It is important for indicator devices to be calibrated before each use.

Rim-and-face There are two variations of the rim-and-face method. One requires one rim reading and one face reading at the stationary machine, where the dial indicator mounting brackets and posts are attached to the machine to be shimmed. The other method is identical, except that the rim and face readings are taken at the machine to be shimmed, where the dial indicator mounting brackets and posts are attached to the stationary machine.

As with the reverse-dial indicator method, the measuring device used for rim-and-face alignment is also a dial indicator. The fixture has two runout indicators mounted on a common arm as opposed to reverse-dial fixtures, which have two runout indicators mounted on two separate arms.

The rim-and-face gauges measure both the offset and angularity for the movable machine train component only (as compared with the reverse-dial method, which measures offset and calculates angularity for both the stationary and movable components). With the rim-and-face method, one dial indicator is mounted perpendicular to the shaft, which defines the offset of the movable shaft. The second indicator is mounted parallel to the shaft, which registers the angularity of the movable shaft. Figure 7.17 illustrates the typical configuration of a rim-and-face fixture.

Mounting As with the reverse-dial alignment fixture, proper mounting of the rim-and-face fixture is essential. The fixture must be rigidly mounted on both the stationary and movable shafts. All mechanical linkages must be tight and looseness held to an absolute minimum. Any fixture movement will distort both the offset and angularity readings as the shafts are rotated through 360 degrees.

Rim-and-face measurements are made in exactly the same manner as those of reverse-dial indicator methods. The shafts are slowly rotated in a clockwise direction in 90-degree increments. Measurements, including positive and negative signs, should be recorded at the 12, 3, 6, and 9 o'clock positions.

Limitations Rim-and-face alignment is subject to the same errors as those of the reverse-dial indicator system, which are discussed in Chapter 3. As with that system, care must be taken to ensure proper orientation with the equipment and accurate recording of the data.

Note that rim-and-face alignment cannot be used when there is any end play, or axial movement, in the shafts of either the stationary or movable machine-train components. Since the dial indicator that is mounted parallel to the shaft is used to measure the angularity of the shafts, any axial movement or "float" in either shaft will distort the measurement.

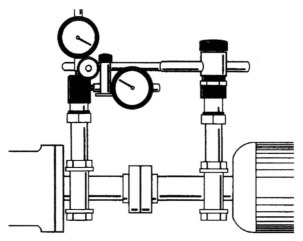

Figure 7.17 Typical configuration of a rim-and-face fixture.

OPTICAL OR LASER ALIGNMENT

Optical or laser alignment systems are based on the same principles as the reverse-dial method but replace the mechanical components such as runout gauges and cantilevered mounting arms with an optical device such as a laser. As with the reverse-dial method, offset is measured and angularity is calculated.

A typical system, which is shown in Figure 7.18, uses two transmitter/sensors rigidly mounted on fixtures similar to the reverse-dial apparatus. When the shaft is rotated to one of the positions of interest (i.e., 12 o'clock, 3 o'clock, etc.), the transmitter projects a laser beam across the coupling. The receiver unit detects the beam, and the offset and angularity are determined and recorded.

Advantages

Optical-alignment systems offer several advantages. Because laser fixtures eliminate the mechanical linkage and runout gauges, there is no fixture sag. This greatly increases the accuracy and repeatability of the data obtained when using this method.

Most of the optical-alignment systems incorporate a microprocessing unit, which eliminates recording errors commonly found with reverse-dial indicator and rim-and-face methods. Optical systems automatically maintain the proper orientation and provide accurate offset and angularity data, virtually eliminating operator error.

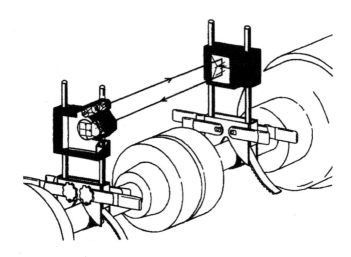

These microprocessor-based systems automatically calculate correction factors. If the fixtures are properly mounted and the shafts are rotated to the correct positions, the system automatically calculates and displays the appropriate correction for each foot of the movable machine-train component. This feature greatly increases the accuracy of the alignment process.

Disadvantages

Since optical-alignment systems are dependent on the transmission of a laser beam, which is a focused beam of light, they are susceptible to problems in some environments. Heat waves, steam, temperature variations, strong sunlight, and dust can distort the beam. When this happens, the system will not perform accurately.

One method that can be used to overcome most of the environment-induced problems is to use plastic tubing to shield the beam. This tubing can be placed between the transmitter and receiver of the optical-alignment fixture. It should be sized to permit transmission and reception of the light beam but small enough to prevent distortion caused by atmospheric or environmental conditions. Typically, 2-inch, thin-wall tubing provides the protection required for most applications.

ALIGNMENT PROCEDURES

This section discusses the procedures for obtaining the measurements needed to align two classes of equipment: (1) horizontally installed units and (2) vertically installed units. The procedures for performing the initial alignment check for offset and angularity and for determining how much correction to make are presented.

Prior to taking alignment measurements, however, remember that it is necessary to remove any soft-foot that is present, making sure that the proper nut-tightening procedure is followed, and to correct for indicator sag (except when using the optical-alignment method). Refer to Chapter 2 for detailed discussions on indicator sag and soft-foot.

Horizontal Units

There are two parts to making alignment measurements on horizontally mounted units, and these are typically taken by using the reverse-dial indicator method. The first part of the procedure is to perform an initial alignment check by obtaining readings for the stationary and movable

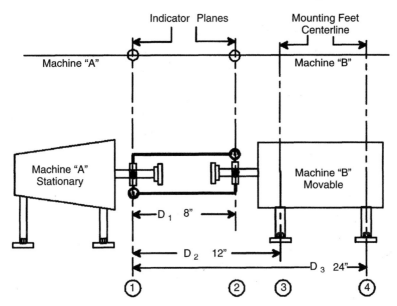

Figure 7.20 Reverse-dial indicator alignment setup.

The vertical and horizontal adjustments necessary to move Machine "B" from the actual position (Figure 7.19 readings) to the desired state of alignment (Figure 7.21 readings) are determined by using the equations below. Note that the desired state of alignment is obtained from manufacturer's tolerances. (When using manufacturer's tolerances, it is important to know if they compensate for thermal growth.)

For example, the shim adjustment at the near foot (N_f) and far foot (F_f) for the readings in Figures 7.19 and 7.21 can be determined by using the vertical movement formulas shown below. Since the top readings equal zero, only the bottom readings are needed in the calculation.

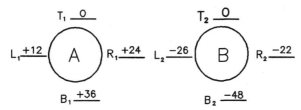

Figure 7.21 Desired dial indicator state readings at ambient conditions.

$$V_1 = \frac{B_3 - B_1}{2} = \frac{(-10) - (+36)}{2} = -23$$

$$V_2 = \frac{B_4 - B_2}{2} + V_1 = \frac{(+20) - (-48)}{2} + (-23) = +11$$

$$N_f = \frac{V_2 \times D_2}{D_1} - V_1 = \frac{(+11) \times (+12)}{8} - (-23) = +40$$

$$F_f = \frac{V_2 \times D_3}{D_1} - V_1 = \frac{(+11) \times (+24)}{8} - (-23) = +56$$

For N_f, at near foot of "B," add 0.040-inch (40 mil) shims. For F_f, at the far foot of "B," add 0.056-inch (56 mil) shims.

For example, the side-to-side movement at N_f and F_f can be determined in the horizontal movement formula:

$$H_1 = \frac{(R_3 - L_3) - (R_1 - L_1)}{2} = \frac{[(-15) - (+5)] - [(+24) - (+12)]}{2} = -16$$

$$H_2 = \frac{(R_4 - L_4) - (R_2 - L_2)}{2} + H_1$$

$$= \frac{[(+6) - (+14)] - [(-22) - (-26)]}{2} + (-16) = -22$$

$$N_f = \frac{H_2 \times D_2}{D_1} - H_1 = \frac{(-22) \times (+12)}{8} - (-16) = -17$$

$$F_f = \frac{H_2 \times D_3}{D_1} - H_1 = \frac{(-22) \times (+24)}{8} - (-16) = -50$$

For N_f, at near foot of "B," move right 0.017 inch.

For F_f, at far foot of "B," move right 0.050 inch.

Vertical Units

The alignment process for most vertical units is quite different from that used for aligning horizontally mounted units. The major reason is that most vertical units are not designed to allow realignment to be performed under the assumption that they will always fit together perfectly. Field checks, however, have proven this assumption to be wrong in a vast majority of cases. Although it is quite difficult to correct misalignment on a vertical unit, it is essential that it be done to increase reliability and decrease maintenance costs.

Initial Alignment Check

The following procedure can be used on vertical units to obtain angularity and offset values needed to compare with recommended manufacturer's (i.e., desired) tolerances to determine if a unit is out of alignment.

- Perform an alignment check on the unit by using the reverse-dial indicator method.
- Install brackets and dial indicators as illustrated in Figure 7.22.
- Check the alignment in two planes by using the following directional designators: "north/south" and "east/west."

Consider the point of reference nearest to you as being "south," which corresponds to the "bottom" position of a horizontal unit. (Note: Indicator sag does not occur when readings are taken as indicated below.)

- Perform the "north/south" alignment checks by setting the indicator dials to "zero" on the "north" side and take the readings on the "south" side.
- Perform the "east/west" alignment checks by setting the indicator dials to "zero" on the "west" side and take the readings on the "east" side.
- Record the distance between the dial indicator centerlines, D_1.
- Record the distance from the centerline of the coupling to the top dial indicator.

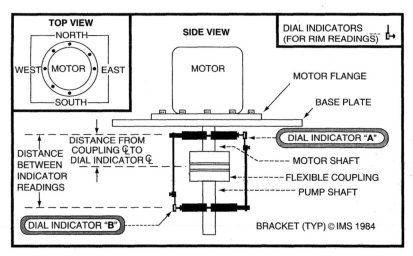

Figure 7.22 Proper dial indicator and bracket positioning when performing a vertical pump alignment.

- Record "zero" for the distance, D_2, from the Indicator A to the "top foot" of the movable unit.
- Record the distance, D_3, from Indicator A to the "bottom foot" of the movable unit.
- Set the top dial indicator to "zero" when it is in the "north" position.

North/South Alignment Check

- Rotate shafts 180 degrees until the top indicator is in the "south" position and obtain a reading.
- Rotate shafts 180 degrees again and check for repeatability of "zero" on the "north" side, then another 180 degrees to check for repeatability of reading obtained on the "south" side.
- Note: If results are not repeatable, check bracket and indicators for looseness and correct as necessary. If repeatable, record the "south" reading.
- Rotate the shafts until the bottom dial indicator is in the "north" position and set it to "zero."
- Rotate the shafts 180 degrees and record "south" side reading. Check for repeatability.

East/West Alignment Check

- Rotate the shafts until the top dial indicator is in the "west" position and set it to "zero."
- Rotate the shafts 180 degrees and obtain the reading on the "east" side. Check for repeatability.
- Rotate the shafts until the bottom dial indicator is in the "west" position and set it to "zero."
- Rotate the shafts 180 degrees and again obtain the reading on the "east" side. Check for repeatability.

Determining Corrections

If the unit must be realigned, with vertical units it is necessary to use the rim-and-face method to obtain offset and angularity readings. Unlike horizontally mounted units, it is not possible to correct both angularity and offset with one adjustment. Instead, we must first correct the angular misalignment in the unit by shimming and then correct the offset by properly positioning the motor base flange on the base plate.

Because most units are designed in such a manner that realignment is not intended, it is necessary to change this design feature. Specifically, the "rabbet fit" between the motor flange and the base plate is the major hindrance to realignment.

Therefore, before proceeding with the alignment method, one should consider that the rabbet fit is designed to automatically "center" the motor during installation. In theory, this should create a condition of perfect alignment between the motor and the driven-unit shafts. The rabbet fit is not designed to support the weight of the unit or resist the torque during start-up or operation; the motor flange and hold-down bolts are designed to do this. Since the rabbet fit is merely a positioning device, it is quite permissible to "bypass" it. This may be accomplished by either of the following:

- Machining off the entire male portion
- Grinding off the male and/or female parts as necessary.

Angularity Correction

There are three steps to follow when correcting for angularity. The first step is to obtain initial readings. The next step is to obtain corrected readings. The third step is to shim the machine.

Step 1: Initial Readings The following procedure is for obtaining initial readings.

- Change the position of the bottom dial indicator so that it can obtain the "face readings" of the lower bracket (see Figure 7.23).

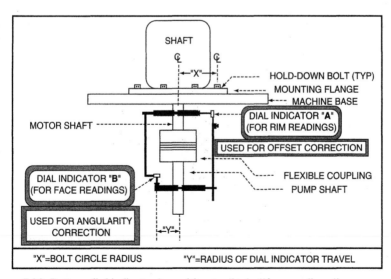

Figure 7.23 Bottom dial indicator in position to obtain "face readings."

- Looking from the "south" side, identify the hold-down bolt at the "north" position and label it #1. Proceeding clockwise, number each hold-down bolt until all are numbered (see Figure 7.24).
- Determine the largest negative reading, which occurs at the widest point, by setting the bottom dial indicator to "zero" at point #1. This should be in line with centerline of hold-down bolt #1. Record the reading.
- Turn the shafts in a clockwise direction and record the data at each hold-down bolt centerline until readings have been taken at all positions.
- Use Figure 7.25 as an example of how the readings are taken. Remember that all readings are taken from the position of looking down on the lower bracket.

Note: We will always be looking for the largest negative (−) reading. If all readings are positive (+), the initial set point of zero will be considered the largest negative (−) reading. In Figure 7.25, the largest negative reading occurs at point #7.

Step 2: Corrected Readings Obtain corrected readings with the following procedure.

- Rotate the shafts until the indicator is again at the point where the largest negative reading occurs.

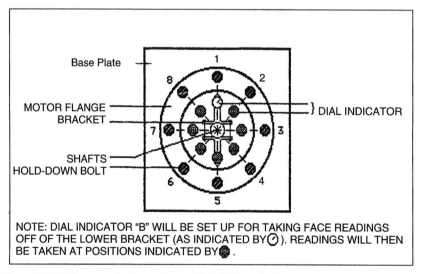

Figure 7.24 Diagram of a base plate with hold-down bolts numbered.

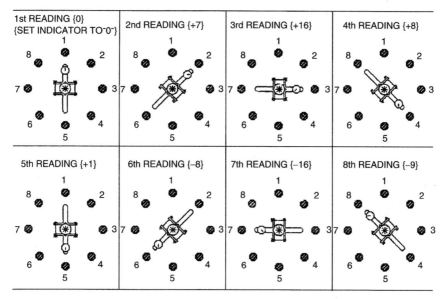

1st READING {0} {SET INDICATOR TO "0"}	2nd READING {+7}	3rd READING {+16}	4th READING {+8}
5th READING {+1}	6th READING {−8}	7th READING {−16}	8th READING {−9}

Figure 7.25 Determining the largest negative reading and the widest point.

- Set the dial indicator to "zero" at this point and take another complete set of readings. With Figure 7.25 as an example, set the dial indicator to "zero" at point #7 (in line with centerline of bolt #7). The results of readings at the other hold-down bolt centerlines are as follows:

#1	+16
#2	+23
#3	+32
#4	+24
#5	+17
#6	+8
#7	0
#8	+7

Step 3: Shimming Perform shimming with the following procedure. Measure the hold-down bolt circle radius and the radius of dial indicator travel as shown in Figure 7.26.

Compute the shim multiplier, X/Y, where:

X = Bolt circle radius

Y = Radius of indicator travel

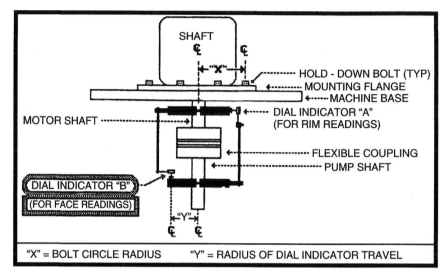

Figure 7.26 Determining bolt circle radius and radius of dial indicator.

For example: If X = 9 inches and Y = 4 inches, the shim multiplier is 9/4 = 2.25. The necessary shimming at each bolt equals the shim multiplier (2.25) times the bolt's corrected reading as determined in Chapter 4.

$$
\begin{aligned}
\#1 \ &- \ 2.25 \times 16 = 36 \ \text{mils} = 0.036 \ \text{inch} \\
\#2 \ &- \ 2.25 \times 23 = 52 \ \text{mils} = 0.052 \ \text{inch} \\
\#3 \ &- \ 2.25 \times 32 = 72 \ \text{mils} = 0.072 \ \text{inch} \\
\#4 \ &- \ 2.25 \times 24 = 54 \ \text{mils} = 0.054 \ \text{inch} \\
\#5 \ &- \ 2.25 \times 17 = 38 \ \text{mils} = 0.038 \ \text{inch} \\
\#6 \ &- \ 2.25 \times 8 \ = 18 \ \text{mils} = 0.018 \ \text{inch} \\
\#7 \ &- \ 2.25 \times 0 \ = 0 \ \ \text{mils} = 0.000 \ \text{inch} \\
\#8 \ &- \ 2.25 \times 7 \ = 16 \ \text{mils} = 0.016 \ \text{inch}
\end{aligned}
$$

Offset Correction Once the angularity has been corrected by making the necessary shim adjustments at each of the hold-down bolts, it is necessary to correct the offset by sliding the movable unit (i.e., motor in this example) on the base plate. The top dial indicator is used to monitor the movements as they are being made. "North/south" and "east/west" designations are used to describe the positioning of the unit.

North/South Correction The following is the procedure for making the "north/south" corrections.

- Rotate shafts until the top dial indicator is in the "north" position. Set it to "zero."
- Rotate the shafts 180 degrees (until the top dial indicator is in the "south" position) and record the reading.
- Determine movement necessary to correct the offset in this plane by dividing the reading by 2. This is the amount of movement (in mils) required. Direction of movement can be determined by the following rule: If the sign of the reading is positive (+), the motor must be moved toward the "north." If negative (−), it must be moved toward the "south."

East/West Correction The following is the procedure for making the "east/west" corrections.

- Rotate the shafts until the top dial indicator is in the "west" position. Set it to "zero."
- Rotate the shafts 180 degrees (until the top dial indicator is in the "east" position) and record the reading.
- Determine movement necessary to correct the offset in this plane by dividing the reading by 2. This value will be the amount of movement (in mils) required. Direction of movement can be determined by the following rule: If the sign of the reading is positive (+), the movable unit (motor) must be moved toward the "west." If negative (−), it must be moved toward the "east."

Making the Offset Corrections After the amounts and directions of required offset adjustments have been obtained, the next step is to actually align the equipment. This is accomplished by using two dial indicators with magnetic bases, which are installed on the south (or north) and west (or east) sides of the mounting flange of the movable unit or motor. See Figure 7.27 for an illustration of this setup. It is important to zero both dial indicators before making adjustments and to watch both dial indicators while moving the unit.

Note: The motor position on the base plate must be adjusted to align the equipment, which may require machining or grinding of the rabbet fit. Remember, however, that the rabbet fit is only a positioning device and is not a structural support.

COMPUTATIONS, ADJUSTMENTS, AND PLOTS

Once initial alignment readings are obtained by using the preceding procedures, they must be adjusted for changes in the machine-train, which can be caused by process movement, vibration, or thermal growth. These adjustments must be

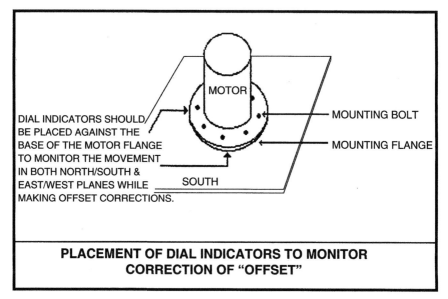

DIAL INDICATORS SHOULD
BE PLACED AGAINST THE
BASE OF THE MOTOR FLANGE
TO MONITOR THE MOVEMENT
IN BOTH NORTH/SOUTH &
EAST/WEST PLANES WHILE
MAKING OFFSET CORRECTIONS.

MOTOR

SOUTH

MOUNTING BOLT

MOUNTING FLANGE

**PLACEMENT OF DIAL INDICATORS TO MONITOR
CORRECTION OF "OFFSET"**

Figure 7.27 Placement of dial indicators to monitor offset corrections.

made to achieve proper alignment at normal operating conditions. Once readings are obtained, the use of graphical plotting helps the technician visualize misalignment and the necessary corrections that must be made and to catch computation errors.

Adjustments for Thermal Growth

Thermal growth generally refers to the expansion of materials with increasing temperature. For alignment purposes, thermal growth is the shaft centerline movement associated with the change in temperature from the alignment process, which is generally performed at ambient conditions, to normal operating conditions. Such a temperature difference causes the elevation of one or both shafts to change and misalignment to result. Temperature changes after alignment produce changes that may affect both offset and angularity of the shafts and can be in the vertical plane, horizontal plane, or any combination.

Proper alignment practices, therefore, must compensate for thermal growth. In effect, the shafts must be misaligned in the ambient condition so they will become aligned when machine temperatures reach their normal operating range. Generally, manufacturers supply dial indicator readings at ambient conditions, which compensate for thermal movement and result in colinear alignment at normal service conditions. When thermal rise information is supplied by the manufacturer or from machine history records, the necessary compensation may be made during the initial alignment procedure.

However, information concerning thermal rise is not available for all equipment. Generally, manufacturers of critical machinery, such as centrifugal air compressors and turbines, will include information relating to thermal rise in their installation manuals in the section dealing with alignment. When this information is not available, the only method to determine the exact amount of compensation necessary to correct for thermal rise is referred to as a "hot alignment check."

Thermal Compensation Targets

A simple procedure for determining thermal compensation targets is to calculate the movement of the shaft due to temperature change at the bearings or feet. Note that calculated thermal growth is highly dependent on the accuracy of the temperature assumptions and is useful only for initial alignment estimates. Therefore the targets developed from the following procedure should be revised when better data become available.

The formula for this calculation is very simple and very accurate. It requires three factors: (1) the difference in temperature of the machine housing between the feet and shaft bearings, (2) the distance between the shaft centerline and the feet, and (3) the coefficient of thermal expansion of the machine housing material. The thermal growth between any two points of any metal can be predicted by the formula:

$$\text{Growth} = \Delta T \times L \times C$$

Where:
ΔT = Temperature difference between the feet and shaft bearings, °F
L = Length between points (often the vertical distance from the shim plane to the shaft centerline), inches
C = Growth factor (coefficient of thermal expansion)
Growth factors (mils/inch/°F) for common materials are:

Aluminum	0.0126
Bronze	0.0100
Cast Iron, Gray	0.0059
Stainless Steel	0.0074
Mild Steel, Ductile Iron	0.0063

Note: The thermal growth formula is usually applied only to the vertical components of the machine. While the formula can be applied to horizontal growth, this direction is often ignored.

For vertical growth, L is usually taken as the vertical height from the bottom of the foot where shims touch the machine to the shaft centerline. In the case where the machine is mounted on a base that has significant temperature variations along its length, L is the vertical distance from the concrete or other constant temperature baseline to the shaft centerline.

Hot Alignment Check

A hot alignment check is performed exactly like an ambient alignment check (see Chapter 4) with the added safety precautions required for hot machinery. The accuracy of a hot alignment check depends on how soon after shutdown dial indicator readings can be taken. Readings may be taken within a few minutes with the use of shaft-mounted brackets that span a flexible coupling. To speed up the process, assemble the brackets to the fullest extent possible prior to shutdown so that they need only be bolted to the shafts once the machine stops rotating.

Adjustments for Sag and Soft-Foot

The procedure for making adjustments to the readings to account for indicator sag is presented in Chapter 2.

Graphical Plotting

The graphical plotting technique for computing initial alignment can be performed with any of the three types of measurement fixtures (i.e., reverse-dial indicator, rim-and-face, or optical). The following steps should be followed when plotting alignment problems:

    1. Determine the following dimensions from the machine-train, which are illustrated in Figure 7.28:

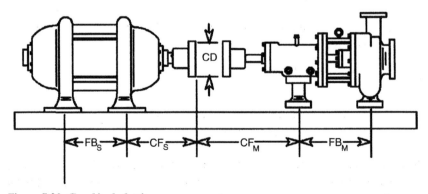

Figure 7.28 Graphical plotting measurements.

FB_S = Front-foot to back-foot of stationary train component
CF_S = Front-foot to coupling of stationary train component
CF_M = Coupling to front-foot of movable train component
FB_M = Front-foot to back-foot of movable train component
CD = Coupling or working diameter

2. On graph or grid paper, pick a horizontal line to be used as the baseline (also referred to as *reference line* or *zero-line*). This line usually crosses the center of the page from left to right and represents the rotational centerline of the stationary machine-train component.
3. Determine the number of inches or mils that each block on the graph paper represents by first finding the distance from the back-foot of the stationary component to the back-foot of the movable component. Then determine the inches or mils per square that will spread the entire machine-train across the graph paper.
4. Plot inches or mils horizontally from left to right.
5. Plot mils from top to bottom vertically. As a general rule, assign 0.5, 1, 2, 5, 10 mils to each vertical step. Note that this scale may need to be changed in cases where excessive misalignment is present.

Known Foot Correction Values

The following steps should be followed to plot misalignment when foot correction values are known (see Figure 7.29):

1. On the baseline, start at the left end and mark the stationary back-foot. From the back-foot and moving right, count the number of

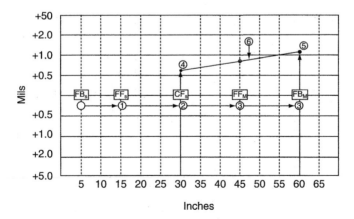

Figure 7.29 Graphical plotting of known foot correction.

squares along the baseline corresponding to FF$_S$. Mark the stationary front-foot location.

2. Starting at the stationary front-foot and moving right, count the number of squares along the baseline corresponding to CF$_S$ and mark the coupling location.

3. Continue this process until the entire machine-train is indicated on the graph.

4. To plot misalignment, locate the CF$_S$ or coupling on the horizontal baseline. From that point, count up or down on the vertical axis until the amount of offset is located on the mils scale. Mark this point on the graph. Use care to ensure that the location is accurately located. Positive values should be above the horizontal baseline and negative values below the line.

5. Locate the FB$_M$ or back-foot of the movable component. Move either up or down vertically on the scale to the point of the offset measurement. Mark this point on the graph. Remember, positive values are above the horizontal baseline and negative values below the line.

6. Draw a line from the back-foot (FB$_M$) of the movable component or MTBM through the front-foot of the movable component toward the vertical line where the stationary coupling is located. Draw a short vertical line at the coupling end of the line. Finish the MTBM by drawing little squares to represent the feet, darkening the line from the back-foot to coupling, and darkening the coupling line.

Known Coupling Results

When plotting coupling misalignment, use the following steps instead of those from the preceding section (see Figure 7.30).

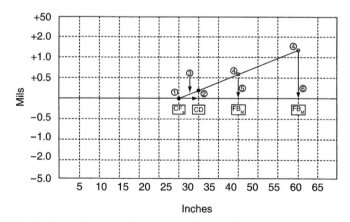

Figure 7.30 Graphical plotting of known coupling results.

1. Start at the stationary coupling location and, moving up or down the vertical axis (mils), count the number of squares corresponding to the vertical or horizontal offset. Move up for positive offset and down for negative offset. Mark a point, which is the MTBM coupling location.
2. Start at the MTBM coupling center and, moving right on the horizontal line, count the number of squares corresponding to the CD dimension (see Figure 7.28) and lightly mark the point. From this point, move up or down vertically on the mils scale the number of squares corresponding to the total mils of angularity per diameter (CD) and mark lightly.
3. From the MTBM coupling center, draw a line through the point marked in the preceding step and extending past the MTBM back-feet location. This line is the MTBM centerline.
4. Now place the MTBM feet. Starting at the MTBM coupling and moving right along a horizontal line, count the number of squares corresponding to CF_M (see Figure 7.28). Then move straight vertically to the MTBM centerline and mark the location of the front-foot. Then starting at the MTBM front-foot and moving right, count the number of squares corresponding to FB_M. From this point, move vertically to the MTBM centerline and mark the location of the MTBM back-foot.
5. Draw a short line perpendicular to the shaft centerline to mark the MTBM coupling. Finish the MTBM by drawing little squares to represent the feet and darkening the line from the back-foot to the coupling.
6. Correction of the MTBM machine-train component can now be measured directly from the graph. Locate the appropriate MTBM foot location and read the actual correction from the vertical or mils scale.

8

ROTOR BALANCING

Mechanical imbalance is one of the most common causes of machinery vibration and is present to some degree on nearly all machines that have rotating parts or rotors. Static, or standing, imbalance is the condition in which there is more weight on one side of a centerline than the other. However, a rotor may be in perfect static balance and not be in a balanced state when rotating at high speed.

If the rotor is a thin disc, careful static balancing may be accurate enough for high speeds. However, if the rotating part is long in proportion to its diameter, and the unbalanced portions are at opposite ends or in different planes, the balancing must counteract the centrifugal force of these heavy parts when they are rotating rapidly.

This section provides information needed to understand and solve the majority of balancing problems by using a vibration/balance analyzer, a portable device that detects the level of imbalance, misalignment, etc., in a rotating part based on the measurement of vibration signals.

SOURCES OF VIBRATION CAUSED BY MECHANICAL IMBALANCE

Two major sources of vibration caused by mechanical imbalance in equipment with rotating parts or rotors are (1) assembly errors and (2) incorrect key length guesses during balancing.

Assembly Errors

Even when parts are precision balanced to extremely close tolerances, vibration caused by mechanical imbalance can be much greater than necessary because of assembly errors. Potential errors include relative placement of each part's center of rotation, location of the shaft relative to the bore, and cocked rotors.

Center of Rotation

Assembly errors are not simply the additive effects of tolerances but also include the relative placement of each part's center of rotation. For example, a "perfectly" balanced blower rotor can be assembled to a "perfectly" balanced shaft and yet the resultant imbalance can be high. This can happen if the rotor is balanced on a balancing shaft that fits the rotor bore within 0.5 mils (0.5 thousandths of an inch) and then is mounted on a standard cold-rolled steel shaft allowing a clearance of over 2 mils.

Shifting any rotor from the rotational center on which it was balanced to the piece of machinery on which it is intended to operate can cause an assembly imbalance four to five times greater than that resulting simply from tolerances. For this reason, all rotors should be balanced on a shaft having a diameter as nearly the same as possible as the shaft on which it will be assembled.

For best results, balance the rotor on its own shaft rather than on a balancing shaft. This may require some rotors to be balanced in an overhung position, a procedure the balancing shop often wishes to avoid. However, it is better to use this technique rather than being forced to make too many balancing shafts. The extra precision balance attained by using this procedure is well worth the effort.

Method of Locating Position of Shaft Relative to Bore

Imbalance often results with rotors that do not incorporate setscrews to locate the shaft relative to the bore (e.g., rotors that are end-clamped). In this case, the balancing shaft is usually horizontal. When the operator slides the rotor on the shaft, gravity causes the rotor's bore to make contact at the 12 o'clock position on the top surface of the shaft. In this position, the rotor is end-clamped in place and then balanced.

If the operator removes the rotor from the balancing shaft without marking the point of bore and shaft contact, it may not be in the same position when reassembled. This often shifts the rotor by several mils as compared with the axis on which it was balanced, thus causing an imbalance to be introduced. The vibration that results is usually enough to spoil what should have been a precision balance and produce a barely acceptable vibration level. In addition, if the resultant vibration is resonant with some part of the machine or structure, a more serious vibration could result.

To prevent this type of error, the balancer operators and those who do final assembly should follow the following procedure. The balancer operator should permanently mark the location of the contact point between the bore and the shaft during balancing. When the equipment is reassembled in the plant or the shop, the assembler should also use this mark. For end-clamped rotors, the assembler should slide the bore on the horizontal shaft, rotating both until the mark is at the 12 o'clock position, and then clamp it in place.

Cocked Rotor

If a rotor is cocked on a shaft in a position different from the one in which it was originally balanced, an imbalanced assembly will result. If, for example, a pulley has a wide face that requires more than one setscrew, it could be mounted on-center but be cocked in a different position than during balancing. This can happen by reversing the order in which the setscrews are tightened against a straight key during final mounting as compared with the order in which the setscrews were tightened on the balancing arbor. This can introduce a pure couple imbalance, which adds to the small couple imbalance already existing in the rotor and causes unnecessary vibration.

For very narrow rotors (i.e., disc-shaped pump impellers or pulleys), the distance between the centrifugal forces of each half may be very small. Nevertheless, a very high centrifugal force, which is mostly counterbalanced statically by its counterpart in the other half of the rotor, can result. If the rotor is slightly cocked, the small axial distance between the two very large centrifugal forces causes an appreciable couple imbalance, which is often several times the allowable tolerance. This is because of the fact that the centrifugal force is proportional to half the rotor weight (at any one time, half of the rotor is pulling against the other half) times the radial distance from the axis of rotation to the center of gravity of that half.

To prevent this, the assembler should tighten each setscrew gradually—first one, then the other, and back again—so that the rotor is aligned evenly. On flange-mounted rotors such as flywheels, it is important to clean the mating surfaces and the bolt holes. Clean bolt holes are important because high couple imbalance can result from the assembly bolt pushing a small amount of dirt between the surfaces, cocking the rotor. Burrs on bolt holes also can produce the same problem.

Other

Other assembly errors can cause vibration. Variances in bolt weights when one bolt is replaced by one of a different length or material can cause vibration. For setscrews that are 90 degrees apart, the tightening sequence may not be the same at final assembly as during balancing. To prevent this, the balancer operator should mark which was tightened first.

Key Length

With a keyed-shaft rotor, the balancing process can introduce machine vibration if the assumed key length is different from the length of the one used during operation. Such an imbalance usually results in a mediocre or "good" running machine as opposed to a very smooth running machine.

For example, a "good" vibration level that can be obtained without following the precautions described in this section is amplitude of 0.12 in./sec (3.0 mm/sec). By following the precautions, the orbit can be reduced to about 0.04 in./sec (1 mm/sec). This smaller orbit results in longer bearing or seal life, which is worth the effort required to make sure that the proper key length is used.

When balancing a keyed-shaft rotor, one half of the key's weight is assumed to be part of the shaft's male portion. The other half is considered to be part of the female portion that is coupled to it. However, when the two rotor parts are sent to a balancing shop for rebalancing, the actual key is rarely included. As a result, the balance operator usually guesses at the key's length, makes up a half key, and then balances the part. (Note: A "half key" is of full-key length but only half-key depth.)

To prevent an imbalance from occurring, do not allow the balance operator to guess the key length. It is strongly suggested that the actual key length be recorded on a tag that is attached to the rotor to be balanced. The tag should be attached in such a way that another device (such as a coupling half, pulley, fan, etc.) cannot be attached until the balance operator removes the tag.

THEORY OF IMBALANCE

Imbalance is the condition in which there is more weight on one side of a centerline than on the other. This condition results in unnecessary vibration, which generally can be corrected by the addition of counterweights. There are four types of imbalance: (1) static, (2) dynamic, (3) couple, and (4) dynamic imbalance combinations of static and couple.

Static

Static imbalance is single-plane imbalance acting through the center of gravity of the rotor, perpendicular to the shaft axis. The imbalance also can be separated into two separate single-plane imbalances, each acting in-phase or at the same angular relationship to each other (i.e., 0 degrees apart). However, the net effect is as if one force is acting through the center of gravity. For a uniform straight cylinder such as a simple paper machine roll or a multi-grooved sheave, the forces of static imbalance measured at each end of the rotor are equal in magnitude (i.e., the ounce-inches or gram-centimeters in one plane are equal to the ounce-inches or gram-centimeters in the other).

In static imbalance, the only force involved is weight. For example, assume that a rotor is perfectly balanced and therefore will not vibrate regardless of the speed of rotation. Also assume that this rotor is placed on frictionless rollers or "knife edges." If a weight is applied on the rim at the center of gravity line between two ends, the weighted portion immediately rolls to the 6 o'clock position because of the gravitational force.

When rotation occurs, static imbalance translates into a centrifugal force. As a result, this type of imbalance is sometimes referred to as "force imbalance," and some balancing machine manufacturers use the word "force" instead of "static" on their machines. However, when the term "force imbalance" was just starting to be accepted as the proper term, an American standardization committee on balancing terminology standardized the term "static" instead of "force." The rationale was that the role of the standardization committee was not to determine and/or correct right or wrong practices but to standardize those currently in use by industry. As a result, the term "static imbalance" is now widely accepted as the international standard and therefore is the term used in this document.

Dynamic

Dynamic imbalance is any imbalance resolved to at least two correction planes (i.e., planes in which a balancing correction is made by adding or removing weight). The imbalance in each of these two planes may be the result of many imbalances in many planes, but the final effects can be characterized to only two planes in almost all situations.

An example of a case in which more than two planes are required is flexible rotors (i.e., long rotors running at high speeds). High speeds are considered to be revolutions per minute (rpm) higher than about 80% of the rotor's first critical speed. However, in over 95% of all run-of-the-mill rotors (e.g., pump impellers, armatures, generators, fans, couplings, pulleys, etc.), two-plane dynamic balance is sufficient. Therefore, flexible rotors are not covered in this document because of the low number in operation and the fact that specially trained people at the manufacturer's plant almost always perform balancing operations.

In dynamic imbalance, the two imbalances do not have to be equal in magnitude to each other, nor do they have to have any particular angular reference to each other. For example, they could be 0 (in-phase), 10, 80, or 180 degrees from each other.

Although the definition of dynamic imbalance covers all two-plane situations, an understanding of the components of dynamic imbalance is needed so that its causes can be understood. Also, an understanding of the components makes it easier to understand why certain types of balancing do not always work with many older balancing machines for overhung rotors and very narrow rotors. The primary components of dynamic imbalance include the number of points of imbalance, the amount of imbalance, the phase relationships, and the rotor speed.

Points of Imbalance

The first consideration of dynamic balancing is the number of imbalance points on the rotor, as there can be more than one point of imbalance within a rotor assembly. This is especially true in rotor assemblies with more than one rotating element, such as a three-rotor fan or multi-stage pump.

Amount of Imbalance

The amplitude of each point of imbalance must be known to resolve dynamic balance problems. Most dynamic balancing machines or in situ balancing instruments are able to isolate and define the specific amount of imbalance at each point on the rotor.

Phase Relationship

The phase relationship of each point of imbalance is the third factor that must be known. Balancing instruments isolate each point of imbalance and determine their phase relationship. Plotting each point of imbalance on a polar plot does this. In simple terms, a polar plot is a circular display of the shaft end. Each point of imbalance is located on the polar plot as a specific radial, ranging from 0 to 360 degrees.

Rotor Speed

Rotor speed is the final factor that must be considered. Most rotating elements are balanced at their normal running speed or over their normal speed range. As a result, they may be out of balance at some speeds that are not included in the balancing solution. As an example, the wheels and tires on your car are dynamically balanced for speeds ranging from zero to the maximum expected speed (i.e., 80 miles per hour). At speeds above 80 miles per hour, they may be out of balance.

COUPLED

Coupled imbalance is caused by two equal non-colinear imbalance forces that oppose each other angularly (i.e., 180 degrees apart). Assume that a rotor with pure couple imbalance is placed on frictionless rollers. Because the imbalance weights or forces are 180 degrees apart and equal, the rotor is statically balanced. However, a pure couple imbalance occurs if this same rotor is revolved at an appreciable speed.

Each weight causes a centrifugal force, which results in a rocking motion or rotor wobble. This condition can be simulated by placing a pencil on a table, then at one end pushing the side of the pencil with one finger. At the same time, push in the opposite direction at the other end. The pencil will tend to rotate end-over-end. This end-over-end action causes two imbalances "orbits," both 180 degrees out of phase, resulting in a "wobble" motion.

Dynamic Imbalance Combinations of Static and Couple

Visualize a rotor that has only one imbalance in a single plane. Also visualize that the plane is not at the rotor's center of gravity but is off to one side. Although there is no other source of couple, this force to one side of the rotor not only causes translation (parallel motion caused by pure static imbalance) but also causes the rotor to rotate or wobble end-over-end as from a couple. In other words, such a force would create a combination of both static and couple imbalance. This again is dynamic imbalance.

In addition, a rotor may have two imbalance forces exactly 180 degrees opposite to each other. However, if the forces are not equal in magnitude, the rotor has a static imbalance in combination with its pure couple. This combination is also dynamic imbalance.

Another way of looking at it is to visualize the usual rendition of dynamic imbalance—imbalance in two separate planes at an angle and magnitude relative to each other not necessarily that of pure static or pure couple.

For example, assume that the angular relationship is 80 degrees and the magnitudes are 8 units in one plane and 3 units in the other. Normally, you would simply balance this rotor on an ordinary two-plane dynamic balancer and that would be satisfactory. But for further understanding of balancing, imagine that this same rotor is placed on static balancing rollers, whereby gravity brings the static imbalance components of this dynamically out-of-balance rotor to the 6 o'clock position.

The static imbalance can be removed by adding counterbalancing weights at the 12 o'clock position. Although statically balanced, the two remaining forces result in a pure couple imbalance. With the entire static imbalance removed, these two forces are equal in magnitude and exactly 180 degrees apart. The couple imbalance can be removed, as with any other couple imbalance, by using a two-plane dynamic balancer and adding counterweights.

Note that whenever you hear the word "imbalance," mentally add the word "dynamic" to it. Then when you hear "dynamic imbalance," mentally visualize "combination of static and couple imbalance." This will be of much help not only in balancing but in understanding phase and coupling misalignment as well.

BALANCING

Imbalance is one of the most common sources of major vibration in machinery. It is the main source in about 40% of the excessive vibration situations. The vibration frequency of imbalance is equal to one times the rpm (1 × rpm) of the imbalanced rotating part.

Before a part can be balanced with the vibration analyzer, certain conditions must be met:

- The vibration must be caused by mechanical imbalance, and
- Weight corrections can be made on the rotating component.

To calculate imbalance units, simply multiply the amount of imbalance by the radius at which it is acting. In other words, 1 ounce. of imbalance at a 1-in. radius will result in 1 oz.-in. of imbalance. Five ounces at a 0.5-in. radius results in 2.5 oz.-in. of imbalance. (Dynamic imbalance units are measured in ounce-inches [oz.-in.] or gram-millimeters [g-mm].) Although this refers to a single plane, dynamic balancing is performed in at least two separate planes. Therefore the tolerance is usually given in single-plane units for each plane of correction.

Important balancing techniques and concepts to be discussed in the sections to follow include in-place balancing, single-plane versus two-plane balancing, precision balancing, techniques that make use of a phase shift, and balancing standards.

IN-PLACE BALANCING

In most cases, weight corrections can be made with the rotor mounted in its normal housing. The process of balancing a part without taking it out of the machine is called *in-place balancing*. This technique eliminates costly and time-consuming disassembly. It also prevents the possibility of damage to the rotor, which can occur during removal, transportation to and from the balancing machine, and reinstallation in the machine.

SINGLE-PLANE VERSUS TWO-PLANE BALANCING

The most common rule of thumb is that a disc-shaped rotating part usually can be balanced in one correction plane only, whereas parts that have appreciable width require two-plane balancing. Precision tolerances, which become more meaningful for higher performance (even on relatively narrow face width), suggest two-plane balancing. However, the width should be the guide, not the diameter-to-width ratio.

For example, a 20-inch-wide rotor could have a large enough couple imbalance component in its dynamic imbalance to require two-plane balancing. (Note: The couple component makes two-plane balancing important.) Yet if the 20-inch width is on a rotor of large enough diameter to qualify as a "disc-shaped rotor," even some of the balance manufacturers erroneously would call for a single-plane balance.

It is true that the narrower the rotor, the less the chance for a large couple component and therefore the greater the possibility of getting by with a single-plane balance. For rotors over 4–5 in. in width, it is best to check for real dynamic imbalance (or for couple imbalance).

Unfortunately, you cannot always get by with a static- and couple-type balance, even for very narrow flywheels used in automobiles. Although most of the flywheels are only 1–1.5 in. wide, more than half have enough couple imbalance to cause excessive vibration. This obviously is not caused by a large distance between the planes (width) but rather by the fact that the flywheel's mounting surface can cause it to be slightly cocked or tilted. Instead of the flywheel being 90 degrees to the shaft axis, it may be perhaps 85 to 95 degrees, causing a large couple despite its narrow width.

This situation is very common with narrow and disc-shaped industrial rotors such as single-stage turbine wheels, narrow fans, and pump impellers. The original manufacturer often accepts the guidelines supplied by others and performs a single-plane balance only. By obtaining separate readings for static and couple, the manufacturer could and should easily remove the remaining couple.

An important point to remember is that static imbalance is always removed first. In static and couple balancing, remove the static imbalance first and then remove the couple.

PRECISION BALANCING

Most original-equipment manufacturers balance to commercial tolerances, a practice that has become acceptable to most buyers. However, because of frequent customer demands, some of the equipment manufacturers now provide precision balancing. Part of the driving force for providing this service is that many large mills and refineries have started doing their own precision balancing to tolerances considerably closer than those used by the original-equipment manufacturer. For example, the International Standards Organization (ISO) for process plant machinery calls for a G6.3 level of balancing in its balancing guide. This was a calculated based on a rotor running free in space with a restraint vibration of 6.3 mm/sec (0.25 in./sec) vibration velocity.

Precision balancing requires a G2.5 guide number, which is based on 2.5 mm/sec (0.1 in./sec) vibration velocity. As can be seen from this, 6.3 mm/sec (0.25 in./sec) balanced rotors will vibrate more than the 2.5 mm/sec (0.1 in./sec) precision balanced rotors. Many vibration guidelines now consider 2.5 mm/sec (0.1 in./sec) "good," creating the demand for precision balancing. Precision balancing tolerances can produce velocities of 0.01 in./sec (0.3 mm/sec) and lower.

It is true that the extra weight of non-rotating parts (i.e., frame and foundation) reduces the vibration somewhat from the free-in-space amplitude. However, it is possible to reach precision balancing levels in only two or three additional runs, providing the smoothest running rotor. The extra effort to the balance operator is minimal because he already has the "feel" of the rotor and has the proper setup and tools in hand. In addition, there is a large financial payoff for this minimal extra effort because of decreased bearing and seal wear.

TECHNIQUES USING PHASE SHIFT

If we assume that there is no other source of vibration other than imbalance (i.e., we have perfect alignment, a perfectly straight shaft, etc.), it is readily seen that pure static imbalance gives in-phase vibrations and pure coupled imbalance gives various phase relationships. Compare the vertical reading of a bearing at one end of the rotor with the vertical reading at the other end of the rotor to determine how that part is shaking vertically. Then compare the horizontal reading at one end with the horizontal reading at the other end to determine how the part is shaking horizontally.

If there is no resonant condition to modify the resultant vibration phase, then the phase for both vertical and horizontal readings is essentially the same, even though the vertical and horizontal amplitudes do not necessarily correspond. In actual practice, this may be slightly off because of other vibration sources such as misalignment. In performing the analysis, what counts is that when the source of the vibration is primarily from imbalance, then the vertical reading phase differences between one end of the rotor and the other will be very similar to the phase differences when measured horizontally. For example, vibrations 60 degrees out of phase vertically would show 60 degrees out of phase horizontally within 20%.

However, the horizontal reading on one bearing will not show the same phase relationship as the vertical reading on the same bearing. This is caused by the pickup axis being oriented in a different angular position as well as the phase adjustment caused by possible resonance. For example, the horizontal vibration frequency may be below the horizontal resonance of various major portions of machinery, whereas the vertical vibration frequency may be above the natural frequency of the floor supporting the machine.

First, determine how the rotor is vibrating vertically by comparing "vertical only" readings with each other. Then determine how the rotor is vibrating horizontally. If the rotor is shaking horizontally and vertically and the phase differences are relatively similar, then the source of vibration is likely to be imbalance. However, before coming to a final conclusion, be sure that other 1 x rpm sources (e.g., bent shaft, eccentric armature, misaligned coupling) are not at fault.

BALANCING STANDARDS

The ISO has published standards for acceptable limits for residual imbalance in various classifications of rotor assemblies. Balancing standards are given in ounce-inches or pound-inches per pound of rotor weight or the equivalent in metric units (gram-millimeters per kilogram). The ounce-inches are for each correction plane for which the imbalance is measured and corrected.

Caution must be exercised when using balancing standards. The recommended levels are for residual imbalance, which is defined as imbalance of any kind that remains after balancing.

Figure 8.1 and Table 8.1 are the norms established for most rotating equipment. Additional information can be obtained from ISO 5406 and 5343. Similar

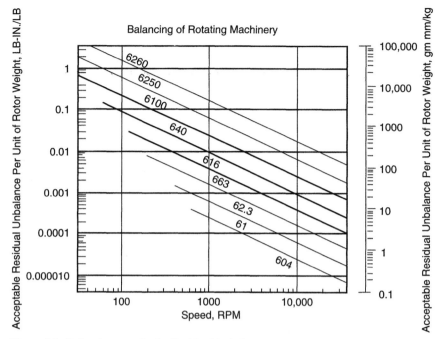

Figure 8.1 Balancing standards. Residual imbalance per unit rotor weight.

standards are available from the American National Standards Institute (ANSI) in their publication ANSI S2.43-1984.

So far, there has been no consideration of the angular positions of the usual two points of imbalance relative to each other or the distance between the two correction planes. For example, if the residual imbalances in each of the two planes were in phase, they would add to each other to create more static imbalance.

Most balancing standards are based on a residual imbalance and do not include multi-plane imbalance. If they are approximately 180 degrees to each other, they form a couple. If the distance between the planes is small, the resulting couple is small; if the distance is large, the couple is large. A couple creates considerably more vibration than when the two residual imbalances are in phase. Unfortunately, there is nothing in the balancing standards that takes this into consideration.

There is another problem that could also result in excessive imbalance-related vibration even when the ISO standards have been met. The ISO standards call for a balancing grade of G6.3 for components such as pump impellers, normal

Table 8.1 Balance Quality Grades For Various Groups Of Rigid Rotors

Balance Quality Grade	Type Of Rotor
G4,000	Crankshaft drives of rigidly mounted slow marine diesel engines with uneven number of cylinders
G1,600	Crankshaft drives of rigidly mounted large two-cycle engines
G630	Crankshaft drives of rigidly mounted large four-cycle engines; crankshaft drives of elastically mounted marine diesel engines
G250	Crankshaft drives of rigidly mounted fast four-cylinder diesel engines
G100	Crankshaft drives of fast diesel engines with six or more cylinders; complete engines (gasoline or diesel) for cars and trucks.
G40	Car wheels, wheel rims, wheel sets, drive shafts; crankshaft drives of elastically mounted fast four-cycle engines (gasoline and diesel) with six or more cylinders; crankshaft drives for engines of cars and trucks
G16	Parts of agricultural machinery; individual components of engines (gasoline or diesel) for cars and trucks
G6.3	Parts or process plant machines; marine main-turbine gears; centrifuge drums; fans; assembled aircraft gas-turbine rotors; fly wheels; pump impellers; machine-tool and general machinery parts; electrical armatures
G2.5	Gas and steam turbines; rigid turbo-generator rotors; rotors; turbo-compressors; machine-tool drives; small electrical armatures; turbine-driven pumps
G1	Tape recorder and phonograph drives; grinding-machine drives
G0.4	Spindles, disks, and armatures of precision grinders; gyroscopes

electric armatures, and parts of process plant machines. This results in an operating speed vibration velocity of 6.3 mm/sec (0.25 in./sec) vibration velocity. However, practice has shown that an acceptable vibration velocity is 0.1 in./sec and the ISO standard of G2.5 is actually required. As a result of these discrepancies, changes in the recommended balancing grade are expected in the future.

9

BEARINGS

A bearing is a machine element that supports a part—such as a shaft—that rotates, slides, or oscillates in or on it. There are two broad classifications of bearings, plain and rolling element (also called anti-friction). Plain bearings are based on sliding motion made possible through the use of a lubricant. Anti-friction bearings are based on rolling motion, which is made possible by balls or other types of rollers. In modern rotor systems operating at relatively high speeds and loads, the proper selection and design of the bearings and bearing-support structure are key factors affecting system life.

TYPES OF MOVEMENT

The type of bearing used in a particular application is determined by the nature of the relative movement and other application constraints. Movement can be grouped into the following categories: rotation about a point, rotation about a line, translation along a line, rotation in a plane, and translation in a plane. These movements can be either continuous or oscillating.

Although many bearings perform more than one function, they can generally be classified based on types of movement, and there are three major classifications of both plain and rolling element bearings: radial, thrust, and guide. Radial bearings support loads that act radially and at right angles to the shaft center line. These loads may be visualized as radiating into or away from a center point like the spokes on a bicycle wheel. Thrust bearings support or resist loads that act axially. These may be described as endwise loads that act parallel to the center line towards the ends of the shaft. This type of bearing prevents lengthwise or axial motion of a rotating shaft.

Guide bearings support and align members having sliding or reciprocating motion. This type of bearing guides a machine element in its lengthwise motion, usually without rotation of the element.

Table 9.1 gives examples of bearings that are suitable for continuous movement; Table 9.2 shows bearings that are appropriate for oscillatory movement only. For the bearings that allow movements in addition to the one listed, the effect on machine design is described in the column, "Effect of the Other Degrees of Freedom." Table 9.3 compares the characteristics, advantages, and disadvantages of plain and rolling element bearings.

ABOUT A POINT (ROTATIONAL)

Continuous movement about a point is rotation, a motion that requires repeated use of accurate surfaces. If the motion is oscillatory rather than continuous, some additional arrangements must be made in which the geometric layout prevents continuous rotation.

ABOUT A LINE (ROTATIONAL)

Continuous movement about a line is also referred to as *rotation,* and the same comments apply as for movement about a point.

ALONG A LINE (TRANSLATIONAL)

Movement along a line is referred to as *translation.* One surface is generally long and continuous, and the moving component is usually supported on a fluid film or rolling contact to achieve an acceptable wear rate. If the translational movement is reciprocation, the application makes repeated use of accurate surfaces, and a variety of economical bearing mechanisms are available.

IN A PLANE (ROTATIONAL/TRANSLATIONAL)

If the movement in a plane is rotational or both rotational and oscillatory, the same comments apply as for movement about a point. If the movement in a plane is translational or both translational and oscillatory, the same comments apply as for movement along a line.

COMMONLY USED BEARING TYPES

As mentioned before, major bearing classifications are plain and rolling element. These types of bearings are discussed in the sections to follow. Table 9.4 is a

Table 9.1 Bearing Selection Guide (Continuous Movement)

Constraint applied to the movement	Examples of arrangements which allow movement only within this constraint	Examples of arrangements which allow this movement but also have other degrees of freedom	Effect of the other degrees of freedom
About a point	Gimbals	Ball on a recessed plate	Ball must be forced into contact with the plate
About a line	Journal bearing with double thrust location	Journal bearing	Simple journal bearing allows free axial movement as well
	Double conical bearing	Screw and nut	Gives some related axial movement as well
		Ball joint or spherical roller bearing	Allows some angular freedom to the line of rotation
Along a line	Crane wheel restrained between two rails	Railway or crane wheel on a track	These arrangements need to be loaded into contact. This is usually done by gravity. Wheels on a single rail or cable need restraint to prevent rotation about the track member
		Pulley wheel on a cable	
		Hovercraft or hoverpad on a track	
In a plane (rotation)	Double thrust bearing	Single thrust bearing	Single thrust bearing must be loaded into contact
In a plane (translation)		Hovercraft or hoverpad	Needs to be loaded into contact usually by gravity

Source: Bearings—A Tribology Handbook, M.J. Neale, Society of Automotive Engineers, Inc., Butterworth Heinemann Ltd., Oxford, Great Britain, 1993.

Table 9.2 Bearing Selection Guide (Oscillatory Movement)

Constraint applied to the movement	Examples of arrangements which allow movement only within this constraint	Examples of arrangemetns which allow this movement but alos have other degrees of freedom	Effect of the other degrees of freedom
About a point	Hookes joint	Cable connection between components	Cable needs to be kept in tension
About a line	Crossed strip flexue pivot	Torsion suspension	A single torsion suspension gives no lateral location
		Knief-edge pivot	Must be loaded into contact
		Rubber bush	Gives some axial and lateral flexibility as well
		Rocker pad	Gives some related translation as well. Must be loaded into contact
Along a line	Crosshead and guide bars	Piston and cylinder	Piston can rotate as well unless it is located by connecting rod
In a plane (rotation)		Rubber ring or disc	Gvies some axial and lateral flexibility as well
In a plane (translation)	Plate between upper and lower guide blocks	Block sliding on a plate	Must be loaded into contact

Source: Bearings—A Tribology Handbook, M.J. Neale, Society of Automotive Engineers, Inc., Butterworth Heinemann Ltd., Oxford, Great Britain, 1993.

Table 9.3 Comparison of Plain and Rolling Element Bearings

Rolling Element	Plain
Assembly on crankshaft is virtually impossible, except with very short or built-up crankshafts	Assembly on crankshaft is no problem as split bearings can be used
Cost relatively high	Cost relatively low
Hardness of shaft unimportant	Hardness of shaft important with harder bearings
Heavier than plain bearings	Lighter than rolling element bearings
Housing requirement not critical	Rigidity and clamping most important housing requirement
Less rigid than plain bearings	More rigid than rolling element bearings
Life limited by material fatigue	Life not generally limited by material fatigue
Lower friction results in lower power consumption	Higher friction causes more power consumption
Lubrication easy to accomplish, the required flow is low except at high speed	Lubrication pressure feed critically important, required flow is large, susceptible to damage by contaminants and interrupted lubricant flow
Noisy operation	Quiet operation
Poor tolerance of shaft deflection	Moderate tolerance of shaft deflection
Poor tolerance of hard dirt particles	Moderate tolerance of dirt particles, depending on hardness of bearing

Requires more overall space:
 	
  Length: Smaller than plain
  Diameter: Larger than plain
	Requires less overall space:
	
	  Length: Larger than rolling element
	  Diameter: Smaller than rolling element

Running Friction:
 	
  Very low at low speeds
  May be high at high speeds
	Running Friction:
	
	  Higher at low speeds
	  Moderate at usual crank speeds

Smaller radial clearance than plain	Larger radial clearance than rolling element

Source: Integrated Systems, Inc.

Table 9.4 Bearing Characteristic Summary

Bearing Type	Description
Plain	See Table 9.3
Lobed	See Radial, Elliptical
Radial or journal	
Cylindrical	Gas lubricated, low-speed applications
Elliptical	Oil lubricated, gear and turbine applications, stiffer and somewhat more stable bearing
Four-axial grooved	Oil lubricated, higher-speed applications than cylindrical
Partial arc	Not a bearing type, but a theoretical component of grooved and lobed bearing configurations
Tilting pad	High-speed applications where hydrodynamic instability and misalignment are common problems
Thrust	Semi-fluid lubrication state, relatively high friction, lower service pressures with multi-collar version, used at low speeds
Rolling element	See Table 9.3. Radial and axial loads, moderate- to high-speed applications
Ball	Higher speed and lighter load applications than roller bearings
Single-row	
Radial non-filling slot	Also referred to as Conrad or deep-groove bearing; sustains combined radial and thrust loads, or thrust loads alone, in either direction, even at high speeds; not self-aligning
Radial filling slot	Handles heavier loads than non-filling slot
Angular contact radial thrust	Radial loads combined with thrust loads, or heavy thrust loads alone; axial deflection must be limited
Ball-thrust	Very high thrust loads in one direction only, no radial loading, cannot be operated at high speeds

Table 9.4 (*continued*)

Bearing Type	Description
Double-row	Heavy radial with minimal bearing deflection and light thrust loads
Double-roll, self-aligning	Moderate radial and limited thrust loads
Roller	Handles heavier loads and shock better than ball bearings, but are more limited in speed than ball bearings
Cylindrical	Heavy radial loads, fairly high speeds, can allow free axial shaft movement
Needle-type cylindrical or barrel	Does not normally support thrust loads, used in space-limited applications, angular mounting of rolls in double-row version tolerates combined axial and thrust loads
Spherical	High radial and moderate-to-heavy thrust loads, usually comes in double-row mounting that is inherently self-aligning
Tapered	Heavy radial and thrust loads; can be preloaded for maximum system rigidity

Source: Integrated Systems, Inc.

bearings characteristics summary. Table 9.5 is a selection guide for bearings operating with continuous rotation and special environmental conditions. Table 9.6 is a selection guide for bearings operating with continuous rotation and special performance requirements. Table 9.7 is a selection guide for oscillating movement and special environment or performance requirements.

PLAIN BEARINGS

All plain bearings also are referred to as *fluid-film bearings*. In addition, radial plain bearings also are commonly referred to as *journal bearings*. Plain bearings are available in a wide variety of types or styles and may be self-contained units or built into a machine assembly. Table 9.8 is a selection guide for radial and thrust plain bearings.

Plain bearings are dependent on maintaining an adequate lubricant film to prevent the bearing and shaft surfaces from coming into contact, which is necessary to prevent premature bearing failure.

Table 9.5 Bearing Selection Guide For Special Environmental Conditions (Continuous Rotation)

Bearing Type	High Temp.	Low Temp.	Vacuum	Wet/ Humid	Dirt/ Dust	External Vibration
Plain, externally pressurized	1 (With gas lubrication)	2	No (Affected by lubricant feed)	2	2 (1 when gas lubricated)	1
Plain, porous metal (oil impregnated)	4 (Lubricant oxidizes)	3 (May have high starting torque)	Possible with special lubricant	2	Seals essential	2
Plain, rubbing (non-metallic)	2 (Up to temp. limit of material)	2	1	2 (Shaft must not corrode)	2 (Seals help)	2
Plain, fluid film	2 (Up to temp. limit of lubricant)	2 (May have high starting torque)	Possible with special lubricant	2	2 (With seals and filtration)	2
Rolling	Consult manufacturer above 150°C	2	3 (With special lubricant)	3 (With seals)	Sealing essential	3 (Consult manufacturers)
Things to watch with all	Effect of thermal expansion on fits	Effect of thermal expansion on fits			Corrosion	
Fretting	Rating: 1–Excellent, 2–Good, 3–Fair, 4–Poor					

Source: Adapted by Integrated Systems, Inc. from Bearings—A Tribology Handbook, M.J. Neale, Society of Automotive Engineers, Inc., Butterworth Heinemann Ltd., Oxford, Great Britain, 1993.

However, this is difficult to achieve, and some contact usually occurs during operation. Material selection plays a critical role in the amount of friction and the resulting seizure and wear that occurs with surface contact. Refer to Chapter 3 for a discussion of common bearing materials. Note that fluid-film bearings do not have the ability to carry the full load of the rotor assembly at any speed and must have turning gear to support the rotor's weight at low speeds.

Table 9.6 Bearing Selection Guide For Particular Performance Requirements (Continuous Rotation)

Bearing Type	Accurate Radial Location	Axial Load Capacity As Well	Low Starting Torque	Silent Running	Standard Parts Available	Simple Lubrication
Plain, externally pressurized	1	No (Need separate thrust bearing)	1	1	No	4 (Need special system)
Plain, fluid film	3	No (Need separate thrust bearing)	2	1	Some	2 (Usually requires circulation system)
Plain, porous metal (oil impregnated)	2	Some	2	1	Yes	1
Plain, rubbing (non-metallic)	4	Some in most instances	4	3	Some	1
Rolling	2	Yes in most instances	1	Usually satisfactory	Yes	2 (When grease lubricated)

Rating: 1–Excellent/very good, 2–Good, 3–Fair, 4–Poor

Source: Adapted by Integrated Systems, Inc. from Bearings—A Tribology Handbook, M.J. Neale, Society of Automotive Engineers, Inc., Butterworth Heinemann Ltd., Oxford, Great Britain, 1993.

Thrust or Fixed

Thrust plain bearings consist of fixed shaft shoulders or collars that rest against flat bearing rings. The lubrication state may be semi-fluid, and friction is relatively high. In multi-collar thrust bearings, allowable service pressures are considerably lower because of the difficulty in distributing the load evenly between several collars. However, thrust ring performance can be improved by introducing tapered grooves. Figure 9.1 shows a mounting half section for a vertical thrust bearing.

Radial or Journal

Plain radial, or journal, bearings also are referred to as *sleeve* or *Babbit* bearings. The most common type is the full journal bearing, which has 360-degree contact

Table 9.7 Bearing Selection Guide For Special Environments Or Performance (Oscillating Movement)

Bearing Type	High Temp.	Low Temp.	Low Friction	Wet/ Humid	Dirt/ Dust	External Vibration
Knife edge pivots	2	2	1	2 (Watch corrosion)	2	4
Plain, porous metal (oil impregnated)	4 (Lubricant oxidizes)	3 (Friction can be high)	2	2	Sealing essential	2
Plain, rubbing	2 (Up to temp. limit of material)	1	2 (With PTFE)	2 (Shaft must not corrode)	2 (Sealing helps)	1
Rolling	Consult manufacturer above 150°C	2	1	2 (With seals)	Sealing essential	4
Rubber bushes	4	4	Elastically stiff	1	1	1
Strip flexures	2	1	1	2 (Watch corrosion)	1	1

Rating: 1–Excellent, 2–Good, 3–Fair, 4–Poor

Source: Adapted by Integrated Systems, Inc. from Bearings—A Tribology Handbook, M.J. Neale, Society of Automotive Engineers, Inc., Butterworth Heinemann Ltd., Oxford, Great Britain, 1993.

with its mating journal. The partial journal bearing has less than 180-degree contact and is used when the load direction is constant. The sections to follow describe the major types of fluid-film journal bearings: plain cylindrical, four-axial groove, elliptical, partial arc, and tilting-pad.

Plain Cylindrical

The plain cylindrical journal bearing (Figure 9.2) is the simplest of all journal bearing types. The performance characteristics of cylindrical bearings are well established, and extensive design information is available. Practically, use of the unmodified cylindrical bearing is generally limited to gas-lubricated bearings and low-speed machinery.

Table 9.8 Plain Bearing Selection Guide

	Journal Bearings	
Characteristics	**Direct Lined**	**Insert Liners**
Accuracy	Dependent on facilities and skill available	Precision components
Quality (Consistency)	Doubtful	Consistent
Cost	Initial cost may be lower	Initial cost may be higher
Ease of Repair	Difficult and costly	Easily done by replacement
Condition upon extensive use	Likely to be weak in fatigue	Ability to sustain higher peak loads
Materials used	Limited to white metals	Extensive range available

	Thrust Bearings	
Characteristic	**Flanged Journal Bearings**	**Separate Thrust Washer**
Cost	Costly to manufacture	Much lower initial cost
Replacement	Involves whole journal/thrust component	Easily replaced without moving journal bearing
Materials used	Thrust face materials limited in larger sizes	Extensive range available
Benefits	Aids assembly on a production line	Aligns itself with the housing

Source: Adapted by Integrated Systems, Inc. from Bearings—A Tribology Handbook, M.J. Neale, Society of Automotive Engineers, Inc., Butterworth Heinemann Ltd., Oxford, Great Britain, 1993.

Four-Axial Groove Bearing

To make the plain cylindrical bearing practical for oil or other liquid lubricants, it is necessary to modify it by the addition of grooves or holes through which the lubricant can be introduced. Sometimes, a single circumferential groove in the middle of the bearing is used. In other cases, one or more axial grooves are provided.

The four-axial groove bearing is the most commonly used oil-lubricated sleeve bearing. The oil is supplied at a nominal gage pressure that ensures an adequate oil flow and some cooling capability. Figure 9.3 illustrates this type of bearing.

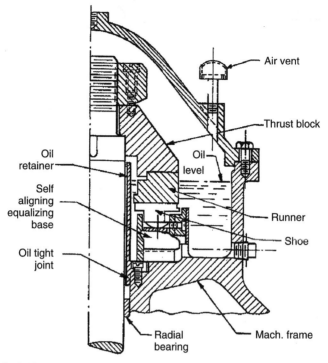

Figure 9.1 Half section of mounting for vertical thrust bearing.

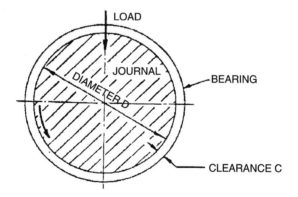

Figure 9.2 Plain cylindrical bearing.

Elliptical Bearing

The elliptical bearing is oil-lubricated and typically is used in gear and turbine applications. It is classified as a lobed bearing in contrast to a grooved bearing. Where the grooved bearing consists of a number of partial arcs with a common center, the lobed bearing is made up of partial arcs whose centers do not

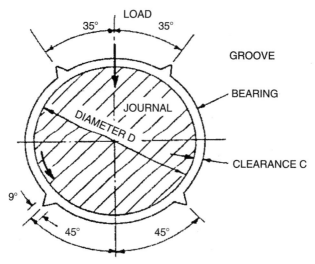

Figure 9.3 Four-axial groove bearing.

coincide. The elliptical bearing consists of two partial arcs in which the bottom arc has its center a distance above the bearing center. This arrangement has the effect of preloading the bearing, where the journal center eccentricity with respect to the loaded arc is increased and never becomes zero. This results in the bearing being stiffened, somewhat improving its stability. An elliptical bearing is shown in Figure 9.4.

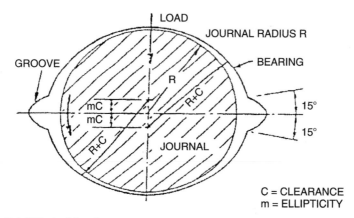

Figure 9.4 Elliptical bearing.

Partial-Arc Bearings

A partial-arc bearing is not a separate type of bearing. Instead, it refers to a variation of previously discussed bearings (e.g., grooved and lobed bearings) that incorporates partial arcs. It is necessary to use partial-arc bearing data to incorporate partial arcs in a variety of grooved and lobed bearing configurations. In all cases, the lubricant is a liquid and the bearing film is laminar. Figure 9.5 illustrates a typical partial-arc bearing.

Tilting-Pad Bearings

Tilting-pad bearings are widely used in high-speed applications in which hydrodynamic instability and misalignment are common problems. This bearing consists of a number of shoes mounted on pivots, with each shoe being a partial-arc bearing. The shoes adjust and follow the motions of the journal, ensuring inherent stability if the inertia of the shoes does not interfere with the adjustment ability of the bearing. The load direction may either pass between the two bottom shoes or it may pass through the pivot of the bottom shoe. The lubricant is incompressible (i.e., liquid) and the lubricant film is laminar. Figure 9.6 illustrates a tilting-pad bearing.

Rolling Element or Anti-Friction

Rolling element anti-friction bearings are one of the most common types used in machinery. Anti-friction bearings are based on rolling motion as opposed to the sliding motion of plain bearings. The use of rolling elements between rotating and stationary surfaces reduces the friction to a fraction of that resulting with the

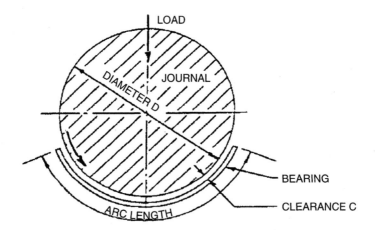

Figure 9.5 Partial-arc bearing.

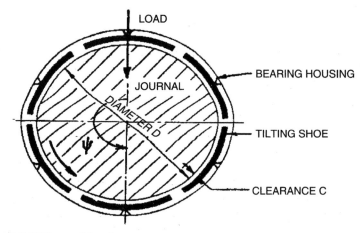

Figure 9.6 Tilting-pad bearing.

use of plain bearings. Use of rolling element bearings is determined by many factors, including load, speed, misalignment sensitivity, space limitations, and desire for precise shaft positioning. They support both radial and axial loads and are generally used in moderate- to high-speed applications.

Unlike fluid-film plain bearings, rolling element bearings have the added ability to carry the full load of the rotor assembly at any speed. Where fluid-film bearings must have turning gear to support the rotor's weight at low speeds, rolling element bearings can maintain the proper shaft centerline through the entire speed range of the machine.

Grade Classifications

Rolling element bearings are available in either commercial- or precision-grade classifications. Most commercial-grade bearings are made to non-specific standards and are not manufactured to the same precise standards as precision-grade bearings. This limits the speeds at which they can operate efficiently, and given brand bearings may or may not be interchangeable.

Precision bearings are used extensively in many machines such as pumps, air compressors, gear drives, electric motors, and gas turbines. The shape of the rolling elements determines the use of the bearing in machinery. Because of standardization in bearing envelope dimensions, precision bearings were once considered to be interchangeable, even if manufactured by different companies. It has been discovered, however, that interchanging bearings is a major cause of machinery failure and should be done with extreme caution.

Rolling Element Types

There are two major classifications of rolling elements: ball and roller. Ball bearings function on point contact and are suited for higher speeds and lighter loads than roller bearings. Roller element bearings function on line contact and gener- ally are more expensive than ball bearings, except for the larger sizes. Roller bearings carry heavy loads and handle shock more satisfactorily than ball bearings but are more limited in speed. Figure 9.7 provides general guidelines to determine if a ball or roller bearing should be selected. This figure is based on a rated life of 30,000 hours.

Although there are many types of rolling elements, each bearing design is based on a series of hardened rolling elements sandwiched between hardened inner and outer rings. The rings provide continuous tracks or races for the rollers or balls to roll in. Each ball or roller is separated from its neighbor by a separator cage or retainer, which properly spaces the rolling elements around the track and guides them through the load zone. Bearing size is usually given in terms of boundary dimensions: outside diameter, bore, and width.

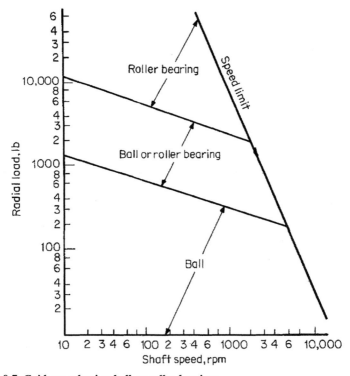

Figure 9.7 Guide to selecting ball or roller bearings.

Ball Bearings

Common functional groupings of ball bearings are radial, thrust, and angular-contact bearings. Radial bearings carry a load in a direction perpendicular to the axis of rotation. Thrust bearings carry only thrust loads, a force parallel to the axis of rotation tending to cause endwise motion of the shaft. Angular-contact bearings support combined radial and thrust loads. These loads are illustrated in Figure 9.8. Another common classification of ball bearings is single row (also referred to as Conrad or deep-groove bearing) and double row.

Single-Row Types of single-row ball bearings are: radial non-filling slot bearings, radial filling slot bearings, angular contact bearings, and ball thrust bearings.

Radial, Non-Filling Slot Bearings This ball bearing is often referred to as the Conrad-type or deep-groove bearing and is the most widely used of all ball bearings (and probably of all anti-friction bearings). It is available in many variations, with single or double shields or seals. They sustain combined radial and thrust loads, or thrust loads alone, in either direction—even at extremely

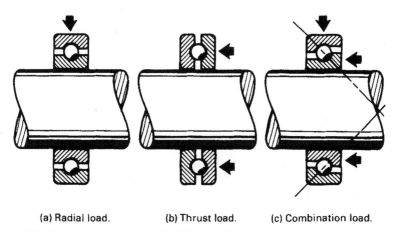

    (a) Radial load.   (b) Thrust load.   (c) Combination load.

Figure 9.8 Three principal types of ball bearing loads.

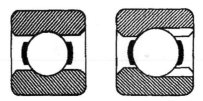

Figure 9.9 Single-row radial, non-filling slot bearing.

high speeds. This bearing is not designed to be self-aligning; therefore, it is imperative that the shaft and the housing bore be accurately aligned.

Figure 9.10 labels the parts of the Conrad anti-friction ball bearing. This design is widely used and is versatile because the deep-grooved raceways permit the rotating balls to rapidly adjust to radial and thrust loadings, or a combination of these loadings.

Radial, Filling Slot Bearing The geometry of this ball bearing is similar to the Conrad bearing, except for the filling slot. This slot allows more balls in the complement and thus can carry heavier radial loads. The bearing is assembled with as many balls that fit in the gap created by eccentrically displacing the inner ring. The balls are evenly spaced by a slight spreading of the rings and heat expansion of the outer ring. However, because of the filling slot, the thrust capacity in both directions is reduced. In combination with radial loads, this bearing design accomodates thrust of less than 60% of the radial load.

Angular Contact Radial Thrust This ball bearing is designed to support radial loads combined with thrust loads, or heavy thrust loads (depending on the contact-angle magnitude). The outer ring is designed with one shoulder higher than the other, which allows it to accommodate thrust loads. The shoulder on the other side of the ring is just high enough to prevent the bearing from separating. This type of bearing is used for pure thrust load in one direction

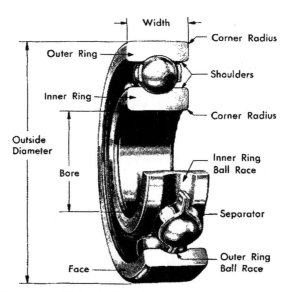

Figure 9.10 Conrad anti-friction ball bearing parts.

and is applied either in opposed pairs (duplex) or one at each end of the shaft. They can be mounted either face-to-face or back-to-back and in tandem for constant thrust in one direction. This bearing is designed for combination loads in which the thrust component is greater than the capacity of single-row, deep-groove bearings. Axial deflection must be confined to very close tolerances.

Ball-Thrust Bearing The ball-thrust bearing supports very high thrust loads in one direction only, but supports no radial loading. To operate successfully, this type of bearing must be at least moderately thrust-loaded at all times. It should not be operated at high speeds, since centrifugal force causes excessive loading of the outer edges of the races.

Double-Row Double-row ball bearings accommodate heavy radial and light thrust loads without increasing the outer diameter of the bearing. However, this type of bearing is approximately 60–80% wider than a comparable single-row bearing. The double-row bearing incorporates a filling slot, which requires the thrust load to be light. Figure 9.11 shows a double-row type ball bearing.

This unit is, in effect, two single-row angular contact bearings built as a unit with the internal fit between balls and raceway fixed during assembly. As a result, fit and internal stiffness are not dependent on mounting methods. These bearings usually have a known amount of internal preload, or compression, built in for maximum resistance to deflection under combined loads with thrust from either direction. As a result of this compression prior to external loading, the bearings are very effective for radial loads in which bearing deflection must be minimized.

Figure 9.11 Double-row type ball bearing.

Another double-row ball bearing is the internal self-aligning type, which is shown in Figure 9.12. It compensates for angular misalignment, which can be caused by errors in mounting, shaft deflection, misalignment, etc. This bearing supports moderate radial loads and limited thrust loads.

Roller As with plain and ball bearings, roller bearings also may be classified by their ability to support radial, thrust, and combination loads. Note that combination load-supporting roller bearings are not called angular-contact bearings as they are with ball bearings. For example, the taper-roller bearing is a combination load-carrying bearing by virtue of the shape of its rollers.

Figure 9.13 shows the different types of roller elements used in these bearings. Roller elements are classified as cylindrical, barrel, spherical, and tapered. Note

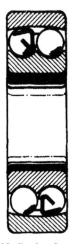

Figure 9.12 Double-row internal self-aligning bearing.

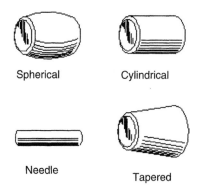

Spherical   Cylindrical

Needle   Tapered

Figure 9.13 Types of roller elements.

that barrel rollers are called needle rollers when less than 0.25-inch in diameter and have a relatively high ratio of length to diameter.

Cylindrical Cylindrical bearings have solid or helically wound hollow cylindrically shaped rollers, which have an approximate length-diameter ratio ranging from 1:1 to 1:3. They normally are used for heavy radial loads beyond the capacities of comparably sized radial ball bearings.

Cylindrical bearings are especially useful for free axial movement of the shaft. The free ring may have a restraining flange to provide some restraint to endwise movement in one direction. Another configuration comes without a flange, which allows the bearing rings to be displaced axially.

Either the rollers or the roller path on the races may be slightly crowned to prevent edge loading under slight shaft misalignment. Low friction makes this bearing type suitable for fairly high speeds. Figure 9.14 shows a typical cylindrical roller bearing.

Figure 9.15 shows separable inner-ring cylindrical roller bearings. Figure 9.16 shows separable inner-ring cylindrical roller bearings with a different inner ring.

The roller assembly in Figure 9.15 is located in the outer ring with retaining rings. The inner ring can be omitted and the roller operated on hardened ground shaft surfaces.

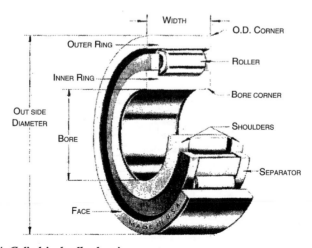

Figure 9.14 Cylindrical roller bearing.

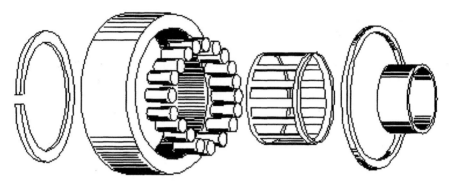

Figure 9.15 Separable inner-ring-type cylindrical roller bearings.

The style in Figure 9.16 is similar to the one in Figure 9.15, except the rib on the inner ring is different. This prohibits the outer ring from moving in a direction toward the rib.

Figure 9.17 shows separable inner-ring-type cylindrical roller bearings with elimination of a retainer ring on one side.

The style shown in Figure 9.17 is similar to the two previous styles except for the elimination of a retainer ring on one side. It can carry small thrust loads in only one direction.

Needle-Type Cylindrical or Barrel Needle-type cylindrical bearings (Figure 9.18) incorporate rollers that are symmetrical with a length at least four times their diameter. They are sometimes referred to as *barrel rollers*. These bearings are most useful where space is limited and thrust-load support is not required. They

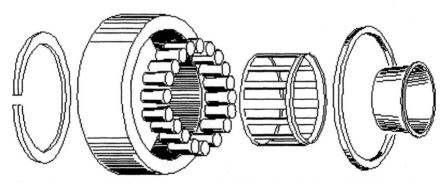

Figure 9.16 Separable inner-ring-type cylindrical roller bearings with different inner ring.

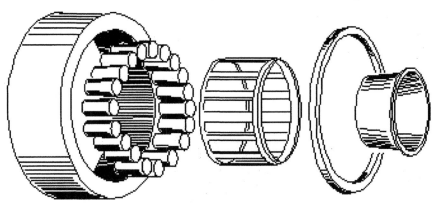

Figure 9.17 Separable inner-ring-type cylindrical roller bearings with elimination of a retainer ring on one side.

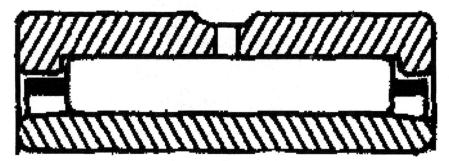

Figure 9.18 Needle bearings.

are available with or without an inner race. If a shaft takes the place of an inner race, it must be hardened and ground. The full-complement type is used for high loads and oscillating or slow speeds. The cage type should be used for rotational motion.

They come in both single-row and double-row mountings. As with all cylindrical roller bearings, the single-row mounting type has a low thrust capacity, but angular mounting of rolls in the double-row type permits its use for combined axial and thrust loads.

Spherical Spherical bearings are usually furnished in a double-row mounting that is inherently self-aligning. Both rows of rollers have a common spherical outer raceway. The rollers are barrel-shaped with one end smaller to provide a small thrust to keep the rollers in contact with the center guide flange.

This type of roller bearing has a high radial and moderate-to-heavy thrust load-carrying capacity. It maintains this capability with some degree of shaft and bearing housing misalignment. While their internal self-aligning feature is useful, care should be taken in specifying this type of bearing to compensate for misalignment. Figure 9.19 shows a typical spherical roller bearing assembly. Figure 9.20 shows a series of spherical roller bearings for a given shaft size.

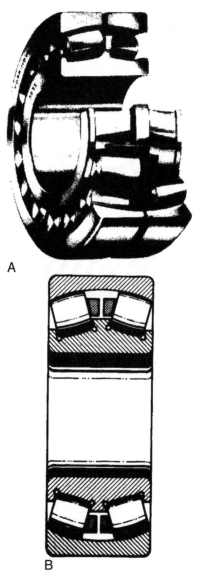

A

B

Figure 9.19 Spherical roller bearing assembly.

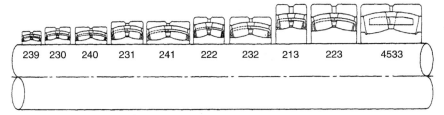

Figure 9.20 Series of spherical roller bearings for a given shaft size (available in several series).

Tapered Tapered bearings are used for heavy radial and thrust loads. They have straight tapered rollers, which are held in accurate alignment by means of a guide flange on the inner ring. Figure 9.21 shows a typical tapered-roller bearing. Figure 9.22 shows necessary information to identify a taper-roller bearing. Figure 9.23 shows various types of tapered roller bearings.

True rolling occurs because they are designed so that all elements in the rolling surface and the raceways intersect at a common point on the axis. The basic characteristic of these bearings is that if the apexes of the tapered working surfaces of both rollers and races were extended, they would coincide on the bearing axis. Where maximum system rigidity is required, they can be adjusted for a preload. These bearings are separable.

BEARING MATERIALS

Because two contacting metal surfaces are in motion in bearing applications, material selection plays a crucial role in their life. Properties of the materials used

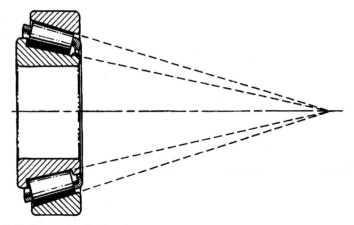

Figure 9.21 Tapered roller bearing.

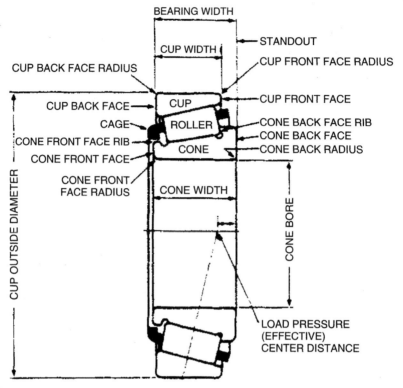

Figure 9.22 Information needed to identify a tapered roller bearing.

in bearing construction determine the amount of sliding friction that occurs, a key factor affecting bearing life. When two similar metals are in contact without the presence of adequate lubrication, friction is generally high and the surfaces will seize (i.e., weld) at relatively low pressures or surface loads. However, certain combinations of materials support substantial loads without seizing or welding as a result of their low frictional qualities.

In most machinery, shafts are made of steel. Bearings are generally made of softer materials that have low frictional as well as sacrificial qualities when in contact with steel. A softer, sacrificial material is used for bearings because it is easier and cheaper to replace a worn bearing as opposed to a worn shaft. Common bearing materials are cast iron, bronze, and babbitt. Other less commonly used materials include wood, plastics, and other synthetics.

There are several important characteristics to consider when specifying bearing materials, including the following: (1) strength or ability to withstand loads

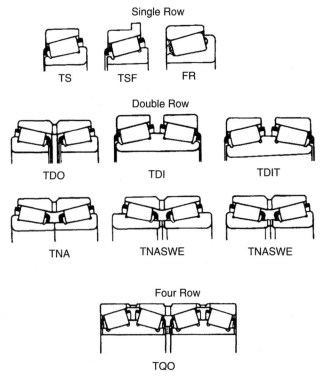

Figure 9.23 Various types of tapered roller bearings.

without plastic deformation; (2) ability to permit embedding of grit or dirt particles that are present in the lubricant; (3) ability to elastically deform to permit load distribution over the full bearing surface; (4) ability to dissipate heat and prevent hot spots that might seize; and (5) corrosion resistance.

PLAIN

As indicated above, dissimilar metals with low frictional characteristics are most suitable for plain bearing applications. With steel shafts, plain bearings made of bronze or babbitt are commonly used. Bronze is one of the harder bearing materials and is generally used for low speeds and heavy loads.

A plain bearing may sometimes be made of a combination of materials. The outer portion may be constructed of bronze, steel, or iron to provide the strength needed to provide a load-carrying capability. The bearing may be lined with a softer material such as babbitt to provide the sacrificial capability needed to protect the shaft.

ROLLING ELEMENT

A specially developed steel alloy is used for an estimated 98% of all rolling element bearing uses. In certain special applications, however, materials such as glass, plastic, and other substances are sometimes used in rolling element construction.

Bearing steel is a high-carbon chrome alloy with high hardenability and good toughness characteristics in the hardened and drawn state. All load-carrying members of most rolling contact bearings are made with this steel.

Controlled procedures and practices are necessary to ensure specification of the proper alloy, maintain material cleanliness, and ensure freedom from defects—all of which affect bearing reliability. Alloying practices that conform to rigid specifications are required to reduce anomalies and inclusions that adversely affect a bearing's useful life. Magnaflux inspections ensure that rolling elements are free from material defects and cracks. Light etching is used between rough and finish grinding processes to stop burning during heavy machining operations.

LUBRICATION

It is critical to consider lubrication requirements when specifying bearings. Factors affecting lubricants include relatively high speeds, difficulty in performing relubrication, non-horizontal shafts, and applications where leakage cannot be tolerated. This section briefly discusses lubrication mechanisms and techniques for bearings.

PLAIN BEARINGS

In plain bearings, the lubricating fluid must be replenished to compensate for end leakage to maintain their load-carrying capacity. Pressure lubrication from a pump- or gravity-fed tank, or automatic lubricating devices such as oil rings or oil disks, are provided in self-contained bearings. Another means of lubrication is to submerge the bearing (in particular, thrust bearings for vertical shafts) in an oil bath.

Lubricating Fluids

Almost any process fluid may be used to lubricate plain bearings if parameters such as viscosity, corrosive action, toxicity, change in state (where a liquid is close to its boiling point), and (in the case of a gaseous fluid) compressibility are appropriate for the application. Fluid-film journal and thrust bearings have run successfully, for example, on water, kerosene, gasoline, acid, liquid refrigerants, mercury, molten metals, and a wide variety of gases.

Gases, however, lack the cooling and boundary-lubrication capabilities of most liquid lubricants. Therefore the operation of self-acting gas bearings is restricted by start/stop friction and wear. If start/stop is performed under load, then the design is limited to about 7 pounds per square inch (lb./in.^{2}) or 48 kilo-Newtons per square meter (kN/m^2) on the projected bearing area, depending on the choice of materials. In general, the materials used for these bearings are those of dry rubbing bearings (e.g., either a hard/hard combination such as ceramics with or without a molecular layer of boundary lubricant or a hard/soft combination with a plastic surface).

Externally pressurized gas journal bearings have the same principle of operation as hydrostatic liquid-lubricated bearings. Any clear gas can be used, but many of the design charts are based on air. There are three forms of external flow restrictors in use with these bearings: pocketed (simple) orifice, unpocketed (annular) orifice, and slot.

State of Lubrication

Fluid or complete lubrication, the condition in which the surfaces are completely separated by a fluid film, provides the lowest friction losses and prevents wear.

The semi-fluid lubrication state exists between the journal and bearing when a load-carrying fluid film does not form to separate the surfaces. This ocurs at comparatively low speed with intermittent or oscillating motion, heavy load, and insufficient oil supply to the bearing. Semi-fluid lubrication also may exist in thrust bearings with fixed parallel-thrust collars; guide bearings of machine tools; bearings with plenty of lubrication, but a bent or misaligned shaft; or where the bearing surface has improperly arranged oil grooves. The coefficient of friction in such bearings may range from 0.02 to 0.08.

In situations where the bearing is well lubricated but the speed of rotation is very slow or the bearing is barely greasy, boundary lubrication takes place. In this situation, which occurs in bearings when the shaft is starting from rest, the coefficient of friction may vary from 0.08 to 0.14.

A bearing may run completely dry in exceptional cases of design or with a complete failure of lubrication. Depending on the contacting surface materials, the coefficient of friction will be between 0.25 and 0.40.

ROLLING ELEMENT BEARINGS

Rolling element bearings also need a lubricant to meet or exceed their rated life. In the absence of high temperatures, however, excellent performance can be obtained with a very small quantity of lubricant. Excess lubricant causes excessive heating, which accelerates lubricant deterioration.

The most popular type of lubrication is the sealed grease ball-bearing cartridge. Grease is commonly used for lubrication because of its convenience and minimum maintenance requirements. A high-quality lithium-based NLGI 2 grease is commonly used for temperatures up to 180°F (82°C). Grease must be replenished and relubrication intervals in hours of operation are dependent on temperature, speed, and bearing size. Table 9.9 is a general guide to the time after which it is advisable to add a small amount of grease.

Some applications, however, cannot use the cartridge design—for example, when the operating environment is too hot for the seals. Another example is when minute leaks or the accumulation of traces of dirt at the lip seals cannot be tolerated (e.g., food processing machines). In these cases, bearings with specialized sealing and lubrication systems must be used.

In applications involving high speed, oil lubrication is typically required. Table 9.10 is a general guide in selecting oil of the proper viscosity for these bearings. For applications involving high-speed shafts, bearing selection must take into

Table 9.9 Ball-Bearing Grease Relubrication Intervals (Hours of Operation)

Bearing Bore, mm	Bearing Speed, rpm				
	5,000	**3,600**	**1,750**	**1,000**	**200**
10	8,700	12,000	25,000	44,000	220,000
20	5,500	8,000	17,000	30,000	150,000
30	4,000	6,000	13,000	24,000	127,000
40	2,800	4,500	11,000	20,000	111,000
50		3,500	9,300	18,000	97,000
60		2,600	8,000	16,000	88,000
70			6,700	14,000	81,000
80			5,700	12,000	75,000
90			4,800	11,000	70,000
100			4,000	10,000	66,000

Source: Marks' Standard Handbook for Mechanical Engineers, 8th Edition, Theodore Baumeister, Ed. McGraw-Hill, New York, 1978.

Table 9.10 Oil Lubrication Viscosity (ISO Identification Numbers)

Bearing Bore, mm	Bearing Speed, rpm				
	10,000	**3,600**	**1,800**	**600**	**50**
4–7	68	150	220		
10–20	32	68	150	220	460
25–45	10	32	68	150	320
50–70	7	22	68	150	320
75–90	3	10	22	68	220
100	3	7	22	68	220

Source: Marks' Standard Handbook for Mechanical Engineers, 8th Edition, Theodore Baumeister, Ed. McGraw-Hill, New York, 1978.

account the inherent speed limitations of certain bearing designs, cooling needs, and lubrication issues such as churning and aeration suppression. A typical case is the effect of cage design and roller-end thrust-flange contact on the lubrication requirements in tapered roller bearings. These design elements limit the speed and the thrust load that these bearings can endure. As a result, it is important to always refer to the bearing manufacturer's instructions on load-carrying design and lubrication specifications.

INSTALLATION AND GENERAL HANDLING PRECAUTIONS

Proper handling and installation practices are crucial to optimal bearing performance and life. In addition to standard handling and installation practices, the issue of emergency bearing substitutions is an area of critical importance. If substitute bearings are used as an emergency means of getting a machine back into production quickly, the substitution should be entered into the historical records for that machine. This documents the temporary change and avoids the possibility of the substitute bearing becoming a permanent replacement. This error can be extremely costly, particularly if the incorrectly specified bearing continually fails prematurely. It is important that an inferior substitute be removed as soon as possible and replaced with the originally specified bearing.

PLAIN BEARING INSTALLATION

It is important to keep plain bearings from shifting sideways during installation and to ensure an axial position that does not interfere with shaft fillets. Both of these can be accomplished with a locating lug at the parting line. Less frequently used is a dowel in the housing, which protrudes partially into a mating hole in the bearing.

The distance across the outside parting edges of a plain bearing are manufactured slightly greater than the housing bore diameter. During installation, a light force is necessary to snap it into place and, once installed, the bearing stays in place because of the pressure against the housing bore.

It is necessary to prevent a bearing from spinning during operation, which can cause a catastrophic failure. Spinning is prevented by what is referred to as "crush." Bearings are slightly longer circumferentially than their mating housings, and on installation, this excess length is elastically deformed or "crushed." This sets up a high radial contact pressure between the bearing and housing, which ensures good back contact for heat conduction and, in combination with the bore-to-bearing friction, prevents spinning. It is important that under no circumstances should the bearing parting lines be filed or otherwise altered to remove the crush.

ROLLER BEARING INSTALLATION

A basic rule of rolling element bearing installation is that one ring must be mounted on its mating shaft or in its housing with an interference fit to prevent rotation. This is necessary because it is virtually impossible to prevent rotation by clamping the ring axially.

Mounting Hardware

Bearings come as separate parts that require mounting hardware or as premounted units that are supplied with their own housings, adapters, and seals.

Bearing Mountings

Typical bearing mountings, which are shown in Figure 9.24, locate and hold the shaft axially and allow for thermal expansion and/or contraction of the shaft. Locating and holding the shaft axially is generally accomplished by clamping one of the bearings on the shaft so that all machine parts remain in proper relationship dimensionally. The inner ring is locked axially relative to the shaft by locating it between a shaft shoulder and some type of removable locking device once the inner ring has a tight fit. Typical removable locking devices are specially designed nuts, which are used for a through shaft, and clamp plates, which are commonly used when the bearing is mounted on the end of the shaft. For the locating or held bearing, the outer ring is clamped axially, usually between housing shoulders or end-cap pilots.

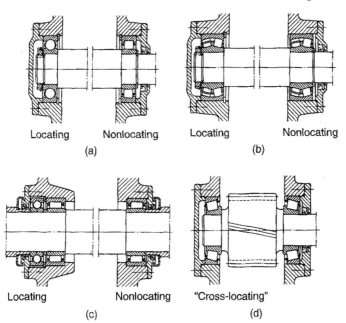

Locating   Nonlocating   Locating   Nonlocating

(a)   (b)

Locating   Nonlocating   "Cross-locating"

(c)   (d)

Figure 9.24 Typical bearing mounting.

With general types of cylindrical roller bearings, shaft expansion is absorbed internally simply by allowing one ring to move relative to the other (Figure 9.24a and 9.24c, non-locating positions). The advantage of this type of mounting is that both inner and outer rings may have a tight fit, which is desirable or even mandatory if significant vibration and/or imbalance exists in addition to the applied load.

Premounted Bearing

Premounted bearings, referred to as *pillow-block* and *flanged-housing* mountings, are of considerable importance to millwrights. They are particularly adaptable to "line-shafting" applications, which are a series of ball and roller bearings supplied with their own housings, adapters, and seals. Premounted bearings come with a wide variety of flange mountings, which permit them to be located on faces parallel or perpendicular to the shaft axis. Figure 9.25 shows a typical pillow block. Figure 9.26 shows a flanged bearing unit.

Inner races can be mounted directly on ground shafts or can be adapter-mounted to "drill-rod" or to commercial shafting. For installations sensitive to imbalance and vibration, the use of accurately ground shaft seats is recommended.

Figure 9.25 Typical pillow block.

Figure 9.26 Flanged bearing unit.

Most pillow-block designs incorporate self-aligning bearing types and do not require the precision mountings utilized with other bearing installations.

Mounting Techniques

When mounting or dismounting a roller bearing, the most important thing to remember is to apply the mounting or dismounting force to the side face of the ring with the interference fit. This force should not pass from one ring to the other through the ball or roller set, because internal damage can easily occur.

Mounting tapered-bore bearings can be accomplished simply by tightening the locknut or clamping plate. This locates it on the shaft until the bearing is forced the proper distance up the taper. This technique requires a significant amount of force, particularly for large bearings.

Cold Mounting

Cold mounting, or force-fitting a bearing onto a shaft or into a housing, is appropriate for all small bearings (i.e., 4-inch bore and smaller). The force, however, must be applied as uniformly as possible around the side face of the bearing and to the ring to be press-fit. Mounting fixtures, such as a simple piece of tubing of appropriate size and a flat plate, should be used. It is not appropriate to use a drift and hammer to force the bearing on, which will cause the bearing to cock. It is possible to apply force by striking the plate with a hammer or by an arbor press. However, before forcing the the bearing on the shaft, a coat of light oil should be applied to the bearing seat on the shaft and the bearing bores. All sealed and shielded ball bearings should be cold mounted in this manner.

Temperature Mounting

The simplest way to mount any open straight-bore bearing regardless of its size is temperature mounting, which entails heating the entire bearing, pushing it on its seat, and holding it in place until it cools enough to grip the shaft. The housing may be heated if practical for tight outside-diameter fits; however, temperatures should not exceed 250°F. If heating of the housing is not practical, the bearing may be cooled with dry ice. The risk of cooling is that if the ambient conditions are humid, moisture is introduced and there is a potential for corrosion in the future. Acceptable ways of heating bearings are by hot plate, temperature-controlled oven, induction heaters, and hot-oil bath.

With the hot plate method, the bearing is simply laid on the plate until it reaches the approved temperature, with a pyrometer or Tempilstik used to make certain it is not overheated. Difficulty in controlling the temperature is the major disadvantage of this method.

When using a temperature-controlled oven, the bearings should be left in the oven long enough to heat thoroughly, but they should never be left overnight.

The use of induction heaters is a quick method of heating bearings. However, some method of measuring the ring temperature (e.g., pyrometer or a Tempilstik) must be used or damage to the bearing may occur. Note that bearings must be demagnetized after the use of this method.

The use of a hot-oil bath is the most practical means of heating larger bearings. Disadvantages are that the temperature of the oil is hard to control, and it may ignite or overheat the bearing. The use of a soluble oil-and-water mixture (10–15% oil) can eliminate these problems and still attain a boiling temperature of 210°F. The bearing should be kept off the bottom of the container by a grate or screen located several inches off the bottom. This is important to allow contaminants to sink to the bottom of the container and away from the bearing.

Dismounting

Commercially available bearing pullers allow rolling element bearings to be dismounted from their seats without damage. When removing a bearing, force should be applied to the ring with the tight fit, although sometimes it is necessary to use supplementary plates or fixtures. An arbor press is equally effective at removing smaller bearings as well as mounting them.

Ball Installation

Figure 9.27 shows the ball installation procedure for roller bearings. The designed load carrying capacity of Conrad-type bearings is determined by the number of balls that can be installed between the rings. Ball installation is accomplished by the following procedure:

- Slip the inner ring slightly to one side
- Insert balls into the gap, which centers the inner ring as the balls are positioned between the rings
- Place stamped retainer rings on either side of the balls before riveting together. This positions the balls equidistant around the bearing.

GENERAL ROLLER-ELEMENT BEARING HANDLING PRECAUTIONS

For roller-element bearings to achieve their design life and perform with no abnormal noise, temperature rise, or shaft excursions, the following precautions should be taken:

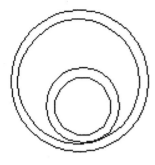

1. The inner ring is moved to one side

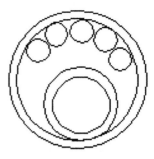

2. Balls are installed in the gap.

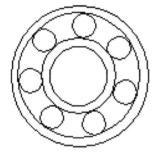

3. The inner ring is centered as the balls are equally positioned in place.

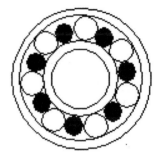

4. A retainer in installed

Figure 9.27 Ball installation procedures.

- Always select the best bearing design for the application and not the cheapest. The cost of the original bearing is usually small by comparison to the costs of replacement components and the down-time in production when premature bearing failure occurs because an inappropriate bearing was used.
- If in doubt about bearings and their uses, consult the manufacturer's representative and the product literature.
- Bearings should always be handled with great care. Never ignore the handling and installation instructions from the manufacturer.
- Always work with clean hands, clean tools, and the cleanest environment available.
- Never wash or wipe bearings prior to installation unless the instructions specifically state that this should be done. Exceptions to this rule are when oil-mist lubrication is to be used and the slushing compound has hardened in storage or is blocking lubrication holes in the bearing rings. In this situation, it is best to clean the bearing with kerosene or

other appropriate petroleum-based solvent. The other exception is if the slushing compound has been contaminated with dirt or foreign matter before mounting.

- Keep new bearings in their greased paper wrappings until they are ready to install. Place unwrapped bearings on clean paper or lint-free cloth if they cannot be kept in their original containers. Wrap bearings in clean, oil-proof paper when not in use.
- Never use wooden mallets, brittle or chipped tools, or dirty fixtures and tools when bearings are being installed.
- Do not spin bearings (particularly dirty ones) with compressed service air.
- Avoid scratching or nicking bearing surfaces. Care must be taken when polishing bearings with emery cloth to avoid scratching.
- Never strike or press on race flanges.
- Always use adapters for mounting that ensure uniform steady pressure rather than hammering on a drift or sleeve. *Never* use brass or bronze drifts to install bearings as these materials chip very easily into minute particles that will quickly damage a bearing.
- Avoid cocking bearings onto shafts during installation.
- Always inspect the mounting surface on the shaft and housing to ensure that there are no burrs or defects.
- When bearings are being removed, clean housings and shafts before exposing the bearings. Dirt is abrasive and detrimental to the designed life span of bearings.
- Always treat used bearings as if they are new, especially if they are to be reused.
- Protect dismantled bearings from moisture and dirt.
- Use clean, filtered, water-free Stoddard's solvent or flushing oil to clean bearings.
- When heating is used to mount bearings onto shafts, follow the manufacturer's instructions.
- When assembling and mounting bearings onto shafts, *never* strike the outer race or press on it to force the inner race. Apply the pressure on the inner race only. When dismantling, follow the same procedure.
- Never press, strike, or otherwise force the seal or shield on factory-sealed bearings.

BEARING FAILURES, DEFICIENCIES, AND THEIR CAUSES

The general classifications of failures and deficiencies requiring bearing removal are overheating, vibration, turning on the shaft, binding of the shaft, noise during operation, and lubricant leakage. Table 9.11 is a troubleshooting guide

Table 9.11 Troubleshooting Guide

Overheating	Vibration	Turning on the Shaft	Binding of the Shaft	Noisy Bearing	Lubricant Leakage
Inadequate or insufficient lubrication	Dirt or chips in bearing	Growth of race due to overheating	Lubricant breakdown	Lubricant breakdown	Overfilling of lubricant
Excessive lubrication	Fatigued race or rolling elements	Fretting wear	Contamination by abrasive or corrosive materials	Inadequate lubrication	Grease churning due to too soft consistency
Grease liquification or aeration	Rotor unbalance	Improper initial fit	Housing distortion or out-of-round pinching bearing	Pinched bearing	Grease deterioration due to excessive operating temperature
Oil foaming	Out-of-round shaft	Excessive shaft deflection	Uneven shimming of housing with loss of clearance	Contamination	Operating beyond grease life
Abrasion or corrosion due to contaminants	Race misalignment	Initial coarse finish on shaft	Tight rubbing seals	Seal rubbing	Seal wear
Housing distortion due to warping or out-of-round	Housing resonance	Seal rub on inner race	Preloaded bearings	Bearing slipping on shaft or in housing	Wrong shaft attitude (bearing seals designed for horizontal mounting only)

(continues)

Table 9.11 (continued)

Overheating	Vibration	Turning on the Shaft	Binding of the Shaft	Noisy Bearing	Lubricant Leakage
Seal rubbing or failure	Cage wear		Cocked races	Flatted roller or ball	Seal failure
Inadequate or blocked scavenge oil passages	Flats on races or rolling elements		Loss of clearance due to excessive adapter tightening	Brinelling due to assembly abuse, handling, or shock loads	Clogged breather
Inadequate bearing clearance or bearing preload	Race turning		Thermal shaft expansion	Variation in size of rolling elements	Oil foaming due to churning or air flow through housing
Race turning	Excessive clearance			Out-of-round or lobular shaft	Gasket (O-ring) failure or misapplication
Cage wear	Corrosion			Housing bore waviness	Porous housing or closure
	False-brinelling or indentation of races			Chips or scores under bearing seat	Lubricator set at the wrong flow rate
	Electrical arcing				
	Mixed rolling element diameters				
	Out-of-square rolling paths in races				

Source: Interrated Systems, Inc.

that lists the common causes for each of these failures and deficiencies. As indicated by the causes of failure listed, bearing failures are rarely caused by the bearing itself.

Many abnormal vibrations generated by actual bearing problems are the result of improper sizing of the bearing liner or improper lubrication. However, numerous machine and process-related problems generate abnormal vibration spectra in bearing data. The primary contributors to abnormal bearing signatures are (1) imbalance, (2) misalignment, (3) rotor instability, (4) excessive or abnormal loads, and (5) mechanical looseness.

Defective bearings that leave the manufacturer are very rare, and it is estimated that defective bearings contribute to only 2% of total failures. The failure is invariably linked to symptoms of misalignment, imbalance, resonance, and lubrication—or the lack of it. Most of the problems that occur result from the following reasons: dirt, shipping damage, storage and handling, poor fit resulting in installation damage, wrong type of bearing design, overloading, improper lubrication practices, misalignment, bent shaft, imbalance, resonance, and soft foot. Any one of these conditions will eventually destroy a bearing—two or more of these problems can result in disaster!

Although most industrial machine designers provide adequate bearings for their equipment, there are some cases where bearings are improperly designed, manufactured, or installed at the factory. Usually, however, the trouble is caused by one or more of the following reasons: (1) improper on-site bearing selection and/or installation, (2) incorrect grooving, (3) unsuitable surface finish, (4) insufficient clearance, (5) faulty relining practices, (6) operating conditions, (7) excessive operating temperature, (8) contaminated oil supply, and (9) oil-film instability.

IMPROPER BEARING SELECTION AND/OR INSTALLATION

There are several things to consider when selecting and installing bearings, including the issue of interchangeability, materials of construction, and damage that might have occurred during shipping, storage, and handling.

Interchangeability

Because of the standardization in envelope dimensions, precision bearings were once regarded as interchangeable among manufacturers. This interchangeability has since been considered a major cause of failures in machinery, and the practice should be used with extreme caution.

Most of the problems with interchangeability stem from selecting and replacing bearings based only on bore size and outside diameters. Often, very little consideration is paid to the number of rolling elements contained in the bearings. This can seriously affect the operational frequency vibrations of the bearing and may generate destructive resonance in the host machine or adjacent machines.

More bearings are destroyed during their installation than fail in operation. Installation with a heavy hammer is the usual method in many plants. Heating the bearing with an oxyacetylene burner is another classical method. However, the bearing does not stand a chance of reaching its life expectancy when either of these installation practices is used. The bearing manufacturer's installation instructions should always be followed.

Materials of Construction

Refer to Chapter 3, which discusses the appropriate materials of construction for the different types of bearings.

Shipping Damage

Bearings and the machinery containing them should be properly packaged to avoid damage during shipping. However, many installed bearings are exposed to vibration, bending, and massive shock loading through bad handling practices during shipping. It has been estimated that approximately 40% of newly received machines have "bad" bearings.

Because of this, all new machinery should be thoroughly inspected for defects before installation. Acceptance criteria should include guidelines that clearly define acceptable design/operational specifications. This practice pays big dividends by increasing productivity and decreasing unscheduled downtime.

Storage and Handling

Stores and other appropriate personnel must be made aware of the potential havoc they can cause by their mishandling of bearings. Bearing failure often starts in the storeroom rather than in the machinery. Premature opening of packages containing bearings should be avoided whenever possible. If packages must be opened for inspection, they should be protected from exposure to harmful dirt sources and then resealed in the original wrappings. The bearing should never be dropped or bumped as this can cause shock loading on the bearing surface.

Incorrect Placement of Oil Grooves

Incorrectly placed oil grooves can cause bearing failure. Locating the grooves in high-pressure areas causes them to act as pressure-relief passages. This interferes with the formation of the hydrodynamic film, resulting in reduced load-carrying capability.

Unsuitable Surface Finish

Smooth surface finishes on both the shaft and the bearing are important to prevent surface variations from penetrating the oil film. Rough surfaces can cause scoring, overheating, and bearing failure. The smoother the finishes, the closer the shaft may approach the bearing without danger of surface contact. Although important in all bearing applications, surface finish is critical with the use of harder bearing materials such as bronze.

Insufficient Clearance

There must be sufficient clearance between the journal and bearing to allow an oil film to form. An average diametral clearance of 0.001 in. per inch of shaft diameter is often used. This value may be adjusted depending on the type of bearing material, the load, speed, and the accuracy of the shaft position desired.

Faulty Relining

Faulty relining occurs primarily with babbitted bearings rather than precision machine-made inserts. Babbitted bearings are fabricated by a pouring process that should be performed under carefully controlled conditions. Some reasons for faulty relining are (1) improper preparation of the bonding surface, (2) poor pouring technique, (3) contamination of babbitt, and (4) pouring bearing to size with journal in place.

Operating Conditions

Abnormal operating conditions or neglecting necessary maintenance precautions cause most bearing failures. Bearings may experience premature and/or catastrophic failure on machines that are operated heavily loaded, speeded up, or being used for a purpose not appropriate for the system design. Improper use of lubricants can also result in bearing failure. Some typical causes of premature failure include (1) excessive operating temperatures, (2) foreign material in the lubricant supply, (3) corrosion, (4) material fatigue, and (5) use of unsuitable lubricants.

Excessive Temperatures

Excessive temperatures affect the strength, hardness, and life of bearing materials. Lower temperatures are required for thick babbitt liners than for thin precision babbitt inserts. Not only do high temperatures affect bearing materials, they also reduce the viscosity of the lubricant and affect the thickness of the film, which affects the bearing's load-carrying capacity. In addition, high temperatures result in more rapid oxidation of the lubricating oil, which can result in unsatisfactory performance.

Dirt and Contamination in Oil Supply

Dirt is one of the biggest culprits in the demise of bearings. Dirt makes its appearance in bearings in many subtle ways, and it can be introduced by bad work habits. It also can be introduced through lubricants that have been exposed to dirt, which is responsible for approximately half of bearing failures throughout the industry.

To combat this problem, soft materials such as babbitt are used when it is known that a bearing will be exposed to abrasive materials. Babbitt metal embeds hard particles, which protects the shaft against abrasion. When harder materials are used in the presence of abrasives, scoring and galling occurs as a result of abrasives caught between the journal and bearing.

In addition to the use of softer bearing materials for applications in which abrasives may potentially be present, it is important to properly maintain filters and breathers, which should regularly be examined. To avoid oil supply contamination, foreign material that collects at the bottom of the bearing sump should be removed on a regular basis.

Oil Film Instability

The primary vibration frequency components associated with fluid-film bearing problems are in fact displays of turbulent or non-uniform oil film. Such instability problems are classified as either oil whirl or oil whip, depending on the severity of the instability.

Machine-trains that use sleeve bearings are designed based on the assumption that rotating elements and shafts operate in a balanced and therefore centered position. Under this assumption, the machine-train shaft will operate with an even, concentric oil film between the shaft and sleeve bearing.

For a normal machine, this assumption is valid after the rotating element has achieved equilibrium. When the forces associated with rotation are in balance, the rotating element will center the shaft within the bearing. However, several problems directly affect this self-centering operation. First, the machine-train must be at designed operating speed and load to achieve equilibrium. Second, any imbalance or abnormal operation limits the machine-train's ability to center itself within the bearing.

A typical example is a steam turbine. A turbine must be supported by auxiliary running gear during start up or shut down to prevent damage to the sleeve bearings. The lower speeds during the start up and shut down phase of operation prevent the self-centering ability of the rotating element. Once the turbine has achieved full speed and load, the rotating element and shaft should operate without assistance in the center of the sleeve bearings.

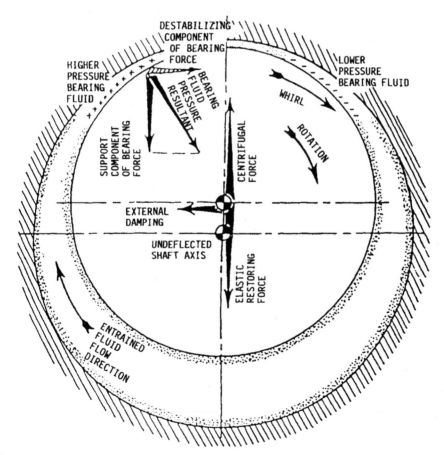

Figure 9.28 Oil whirl, oil whip.

Oil Whirl

In an abnormal mode of operation, the rotating shaft may not hold the centerline of the sleeve bearing. When this happens, an instability called *oil whirl* occurs. Oil whirl is an imbalance in the hydraulic forces within a sleeve bearing. Under normal operation, the hydraulic forces such as velocity and pressure are balanced. If the rotating shaft is offset from the true centerline of the bearing, instability occurs.

As Figure 9.28 illustrates, a restriction is created by the offset. This restriction creates a high pressure and another force vector in the direction of rotation. Oil whirl accelerates the wear and failure of the bearing and bearing support structure.

Oil Whip

The most severe damage results if the oil whirl is allowed to degrade into oil whip. Oil whip occurs when the clearance between the rotating shaft and sleeve bearing is allowed to close to a point approaching actual metal-to-metal contact. When the clearance between the shaft and bearing approaches contact, the oil film is no longer free to flow between the shaft and bearing. As a result, the oil film is forced to change directions. When this occurs, the high-pressure area created in the region behind the shaft is greatly increased. This vortex of oil increases the abnormal force vector created by the offset and rotational force to the point that metal-to-metal contact between the shaft and bearing occurs. In almost all instances where oil whip is allowed, severe damage to the sleeve bearing occurs.

10

COUPLINGS

Couplings are designed to provide two functions: (1) to transmit torsional power between a power source and driven unit, and (2) to absorb torsional variations in the drive train. They are not designed to correct misalignment between two shafts. While certain types of couplings provide some correction for slight misalignment, reliance on these devices to obtain alignment is not recommended.

COUPLING TYPES

The sections that follow provide overviews of the more common coupling types, rigid and flexible. Also discussed are couplings used for special applications, floating-shaft (spacer) and fluid (hydraulic).

Rigid Couplings

A rigid coupling permits neither axial nor radial relative motion between the shafts of the driver and driven unit. When the two shafts are connected solidly and properly, they operate as a single shaft. A rigid coupling is primarily used for vertical applications (e.g., vertical pump). Types of rigid couplings discussed in this section are flanged, split, and compression.

Flanged couplings are used where there is free access to both shafts. Split couplings are used where access is limited on one side. Both flanged and split couplings require the use of keys and keyways. Compression couplings are used when it is not possible to use keys and keyways.

Flanged Couplings

A flanged rigid coupling is composed of two halves, one located on the end of the driver shaft and the other on the end of the driven shaft. These halves are bolted together to form a solid connection. To positively transmit torque, the coupling incorporates axially fitted keys and split circular key rings or dowels, which eliminate frictional dependency for transmission. The use of flanged couplings is restricted primarily to vertical pump shafts. A typical flanged rigid coupling is illustrated in Figure 10.1.

Split Couplings

A split rigid coupling, also referred to as a *clamp* coupling, is basically a sleeve that is split horizontally along the shaft and held together with bolts. It is clamped over the adjoining ends of the driver and driven shafts, forming a solid connection. Clamp couplings are used primarily on vertical pump shafting. A typical split rigid coupling is illustrated in Figure 10.2. As with the flanged coupling, the split rigid coupling incorporates axially fitted keys and split circular key rings to eliminate frictional dependency in the transmission of torque.

Compression Coupling

A rigid compression coupling is composed of three pieces: a compressible core and two encompassing coupling halves that apply force to the core. The core is composed of a slotted bushing that has been machine-bored to fit both ends of the shafts. It also has been machined with a taper on its external diameter from the center outward to both ends. The coupling halves are finish-bored to fit this taper. When the coupling halves are bolted together, the core is compressed down on the shaft by the two halves, and the resulting frictional grip transmits the torque without the use of keys. A typical compression coupling is illustrated in Figure 10.3.

Flexible Couplings

Flexible couplings—which are classified as mechanical flexing, material flexing, or combination—allow the coupled shafts to slide or move relative to each other. Although clearances are provided to permit movement within specified tolerance limits, flexible couplings are not designed to compensate for major misalignments. (Shafts must be aligned to less than 0.002 in. for proper operation.) Significant misalignment creates a whipping movement of the shaft, adds thrust to the shaft and bearings, causes axial vibrations, and leads to premature wear or failure of equipment.

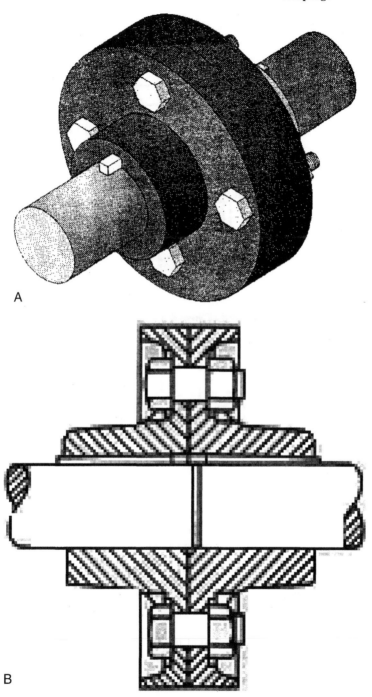

A

B

Figure 10.1 Typical flanged rigid coupling.

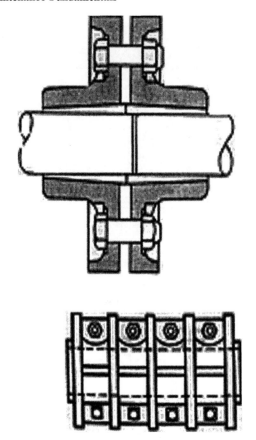

Figure 10.2 Typical split rigid coupling.

Mechanical Flexing

Mechanical-flexing couplings provide a flexible connection by permitting the coupling components to move or slide relative to each other. To permit such movement, clearance must be provided within specified limits. It is important to keep cross loading on the connected shafts at a minimum. This is accomplished by providing adequate lubrication to reduce wear on the coupling components. The most popular of the mechanical-flexing type are the chain and gear couplings.

Chain Chain couplings provide a good means of transmitting proportionately high torque at low speeds. Minor shaft misalignment is compensated for by means of clearances between the chain and sprocket teeth and the clearance that exists within the chain itself.

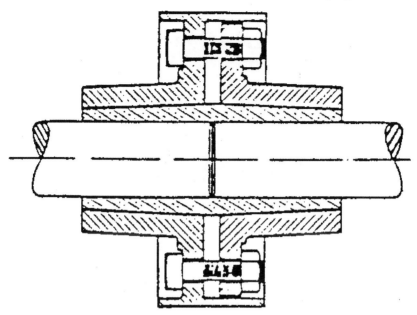

Figure 10.3 Typical compression rigid coupling.

The design consists of two hubs with sprocket teeth connected by a chain of the single-roller, double-roller, or silent type. A typical example of a chain coupling is illustrated in Figure 10.4.

Special-purpose components may be specified when enhanced flexibility and reduced wear are required. Hardened sprocket teeth, special tooth design, and barrel-shaped rollers are available for special needs. Light-duty drives are sometimes supplied with non-metallic chains on which no lubrication should be used.

Gear Gear couplings are capable of transmitting proportionately high torque at both high and low speeds. The most common type of gear coupling consists of two identical hubs with external gear teeth and a sleeve, or cover, with matching internal gear teeth. Torque is transmitted through the gear teeth, whereas the necessary sliding action and ability for slight adjustments in position comes from a certain freedom of action provided between the two sets of teeth.

Slight shaft misalignment is compensated for by the clearance between the matching gear teeth. However, any degree of misalignment decreases the useful life of the coupling and may cause damage to other machine-train components such as bearings. A typical example of a gear-tooth coupling is illustrated in Figure 10.5.

Roller-chain Coupling

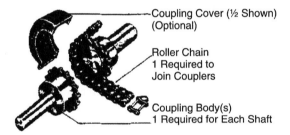

Coupling Cover (½ Shown)
(Optional)

Roller Chain
1 Required to
Join Couplers

Coupling Body(s)
1 Required for Each Shaft

Figure 10.4 Typical chain coupling.

Material Flexing

Material-flexing couplings incorporate elements that accommodate a certain amount of bending or flexing. The material-flexing group includes laminated disk-ring, bellows, flexible shaft, diaphragm, and elastomeric couplings.

Various materials such as metal, plastic, or rubber are used to make the flexing elements in these couplings. The use of the couplings is governed by the operational fatigue limits of these materials. Practically all metals have fatigue limits that are predictable; therefore, they permit definite boundaries of operation to be established. Elastomers such as plastic or rubber, however, usually do not have a well-defined fatigue limit. Their service life is determined primarily by conditions of installation and operation.

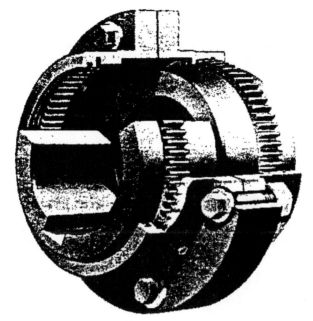

Figure 10.5 Typical gear-tooth coupling.

Laminated Disk-Ring The laminated disk-ring coupling consists of shaft hubs connected to a single flexible disk, or a series of disks, that allows axial movement. The laminated disk-ring coupling also reduces heat and axial vibration that can transmit between the driver and driven unit. Figure 10.6 illustrates some typical laminated disk-ring couplings.

Bellows Bellows couplings consist of two shaft hubs connected to a flexible bellows. This design, which compensates for minor misalignment, is used at moderate rotational torque and shaft speed. This type of coupling provides flexibility to compensate for axial movement and misalignment caused by thermal expansion of the equipment components. Figure 10.7 illustrates a typical bellows coupling.

Flexible Shaft or Spring Flexible shaft or spring couplings are generally used in small equipment applications that do not experience high torque loads. Figure 10.8 illustrates a typical flexible shaft coupling.

Diaphragm Diaphragm couplings provide torsional stiffness while allowing flexibility in axial movement. Typical construction consists of shaft hub flanges

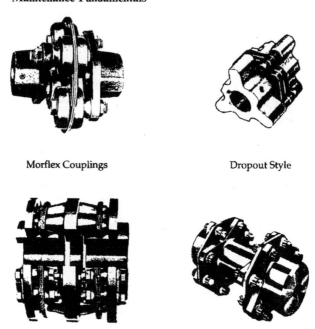

Morflex Couplings   Dropout Style

Laminated disk-ring coupling   Laminated disk-ring coupling
(standard double-engagement)   (high speed spacer type)

Figure 10.6 Typical laminated disk-ring couplings.

Figure 10.7 Typical bellows coupling.

Figure 10.8 Typical flexible shaft coupling.

and a diaphragm spool, which provides the connection between the driver and driven unit. The diaphragm spool normally consists of a center shaft fastened to the inner diameter of a diaphragm on each end of the spool shaft. The shaft hub flanges are fastened to the outer diameter of the diaphragms to complete the mechanical connection. A typical diaphragm coupling is illustrated in Figure 10.9.

Elastomeric Elastomeric couplings consist of two hubs connected by an elastomeric element. The couplings fall into two basic categories, one with the element placed in shear and the other with its element placed in compression. The coupling compensates for minor misalignments because of the flexing capability

Figure 10.9 Typical diaphragm coupling.

of the elastomer. These couplings are usually applied in light- or medium-duty applications running at moderate speeds.

With the shear-type coupling, the elastomeric element may be clamped or bonded in place, or fitted securely to the hubs. The compression-type couplings may be fitted with projecting pins, bolts, or lugs to connect the components. Polyurethane, rubber, neoprene, or cloth and fiber materials are used in the manufacture of these elements.

Although elastomeric couplings are practically maintenance free, it is good practice to periodically inspect the condition of the elastomer and the alignment of the equipment. If the element shows signs of defects or wear, it should be replaced and the equipment realigned to the manufacturer's specifications. Typical elastomeric couplings are illustrated in Figure 10.10.

Combination (Metallic-Grid)

The metallic-grid coupling is an example of a combination of mechanical-flexing and material-flexing type couplings. Typical metallic-grid couplings are illustrated in Figure 10.11.

The metallic-grid coupling is a compact unit capable of transmitting high torque at moderate speeds. The construction of the coupling consists of two flanged hubs, each with specially grooved slots cut axially on the outer edges of the hub flanges. The flanges are connected by means of a serpentine-shaped spring grid that fits into the grooved slots. The flexibility of this grid provides torsional resilience.

Special Application Couplings

Two special application couplings are discussed in this section: (1) the floating-shaft or spacer coupling and (2) the hydraulic or fluid coupling.

Floating-Shaft or Spacer Coupling

Regular flexible couplings connect the driver and driven shafts with relatively close ends and are suitable for limited misalignment. However, allowances sometimes have to be made to accommodate greater misalignment or when the ends of the driver and driven shafts have to be separated by a considerable distance.

Such is the case, for example, with end-suction pump designs in which the power unit of the pump assembly is removed for maintenance by being axially moved toward the driver. If neither the pump nor the driver can be readily removed, they should be separated sufficiently to permit withdrawal of the pump's power unit. An easily removable flexible coupling of sufficient length (i.e., floating-shaft or

Figure 10.10 Typical elastomeric couplings.

spacer coupling) is required for this type of maintenance. Examples of couplings for this type of application are shown in Figure 10.12.

In addition to the maintenance application described above, this coupling (also referred to as *extension* or *spacer sleeve* coupling) is commonly used where equipment is subject to thermal expansion and possible misalignment because of high process temperatures. The purpose of this type of coupling is to prevent harmful misalignment with minimum separation of the driver and driven shaft ends. An example of a typical floating-shaft coupling for this application is shown in Figure 10.13.

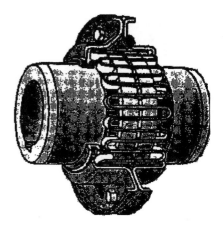

Figure 10.11 Typical metallic-grid couplings.

Laminated disk-ring coupling, spacer type

Gear coupling, spindle type   Gear coupling, high speed spacer type

Figure 10.12 Typical floating-shaft or spacer couplings.

The floating-shaft coupling consists of two support elements connected by a shaft. Manufacturers use various approaches in their designs for these couplings. For example, each of the two support elements may be of the single-engagement type, may consist of a flexible half-coupling on one end and a rigid half-coupling on the other end, or may be completely flexible with some piloting or guiding supports.

Floating-shaft gear couplings usually consist of a standard coupling with a two-piece sleeve. The sleeve halves are bolted to rigid flanges to form two single-flex

Figure 10.13 Typical floating-shaft or spacer couplings for high-temperature applications.

couplings. An intermediate shaft, which permits the transmission of power between widely separated drive components, connects these.

Hydraulic or Fluid

Hydraulic couplings provide a soft start with gradual acceleration and limited maximum torque for fixed operating speeds. Hydraulic couplings are typically used in applications that undergo torsional shock from sudden changes in equipment loads (e.g., compressors). Figure 10.14 is an illustration of a typical hydraulic coupling.

COUPLING SELECTION

Periodically, worn or broken couplings must be replaced. One of the most important steps in performing this maintenance procedure is to ensure that the correct replacement parts are used. After having determined the cause of failure, it is crucial to identify the correct type and size of coupling needed. Even if

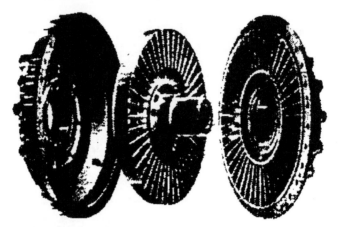

Figure 10.14 Typical hydraulic coupling.

practically identical in appearance to the original, a part still may not be an adequate replacement.

The manufacturer's specification number usually provides the information needed for part selection. If the part is not in stock, a cross-reference guide will provide the information needed to verify ratings and to identify a coupling that meets the same requirements as the original.

Criteria that must be considered in part selection include equipment type, mode of operation, and cost. Each of these criteria is discussed in the sections to follow.

Equipment Type

Coupling selection should be application specific, and therefore it is important to consider the type of equipment that it connects. For example, demanding applications such as variable, high-torque machine-trains require couplings that are specifically designed to absorb radical changes in speed and torque (e.g., metallic-grid). Less demanding applications such as run-out table rolls can generally get by with elastomeric couplings. Table 10.1 lists the coupling type commonly used in a particular application.

Mode of Operation

Coupling selection is highly dependent on the mode of operation, which includes torsional characteristics, speed, and the operating envelope.

Table 10.1 Coupling Application Overview

Application	Coupling* Selection Recommendation
Limited Misalignment Compensation	
Variable, high-torque machine-trains operating at moderate speeds	Metallic-grid combination couplings
Run-out table rolls	Elastomeric flexible couplings
Vertical pump shafting	Flanged rigid couplings, split rigid or clamp couplings
Keys and keyways not appropriate (e.g., brass shafts)	Rigid compression couplings
Transmission of proportionately high torque at low speeds	Chain couplings (mechanical-flexing)
Transmission of proportionately high torque at both high and low speeds	Gear couplings (mechanical-flexing)
Allowance for axial movement and reduction of heat and axial vibration	Laminated disk-ring couplings (material-flexing)
Moderate rotational torque and shaft speed	Bellows couplings (material-flexing)
Small equipment that does not experience high torque loads	Flexible shaft or spring couplings (material-flexing)
Torsional stiffness while allowing flexibility in axial movement	Diaphragm material-flexing couplings
Light- or medium-duty applications running at moderate speeds	Elastomeric couplings (material-flexing)
Gradual acceleration and limited maximum torque for fixed operating speeds (e.g., compressors).	Hydraulic or fluid couplings
Variable or high torque and/or speed transmission	Flexible couplings rated for the maximum torque requirement

(continues)

Table 10.1 (continued)

Application	Coupling* Selection Recommendation
Greater Misalignment Compensation	
Maintenance requiring considerable distance between the driver and driven shaft ends	Floating-shaft or spacer couplings
Misalignment results from expansion due to high process temperatures	

*See Table 10.6 for an application overview for clutches.

Note: Rigid couplings are not designed to absorb variations in torque and speed and should not be used in such applications. Maximum in-service coupling speed should be at least 15% below the maximum coupling speed rating.

Torsional Characteristics

Torque requirements are a primary concern during the selection process. In all applications in which variable or high torque is transmitted from the driver to the driven unit, a flexible coupling rated for the maximum torque requirement must be used. Rigid couplings are not designed to absorb variations in torque and should not be used.

Speed

Two speed-related factors should be considered as part of the selection process: maximum speed and speed variation.

Maximum Speed When selecting coupling type and size, the maximum speed rating must be considered, which can be determined from the vendor's catalog. The maximum in-service speed of a coupling should be well below (at least 15%) the maximum speed rating. The 15% margin provides a service factor that should be sufficient to prevent coupling damage or catastrophic failure.

Speed Variation Variation in speed equates to a corresponding variation in torque. Most variable-speed applications require some type of flexible coupling capable of absorbing these torsional variations.

Operating Envelope

The operating envelope defines the physical requirements, dimensions, and type of coupling needed in a specific application. The envelope information should include shaft sizes, orientation of shafts, required horsepower, full range of operating torque, speed ramp rates, and any other data that would directly or indirectly affect the coupling.

Cost

Coupling cost should not be the deciding factor in the selection process, although it will certainly play a part in it. Although higher-performance couplings may be more expensive, they actually may be the cost-effective solution in a particular application. Selecting the most appropriate coupling for an application not only extends coupling life but also improves the overall performance of the machine-train and its reliability.

INSTALLATION

Couplings must be installed properly if they are to operate satisfactorily. This section discusses shaft and coupling preparation, coupling installation, and alignment.

Shaft Preparation

A careful inspection of both shaft ends must be made to ensure that no burrs, nicks, or scratches are present that will damage the hubs. Potentially damaging conditions must be corrected before coupling installation. Emery cloth should be used to remove any burrs, scratches, or oxidation that may be present. A light film of oil should be applied to the shafts prior to installation.

Keys and keyways also should be checked for similar defects and to ensure that the keys fit properly. Properly sized key stock must be used with all keyways; do not use bar stock or other material.

Coupling Preparation

The coupling must be disassembled and inspected prior to installation. The location and position of each component should be noted so that it can be reinstalled in the correct order. When old couplings are removed for inspection, bolts and bolt holes should be numbered so that they can be installed in the same location when the coupling is returned to service.

Any defects, such as burrs, should be corrected before the coupling is installed. Defects on the mating parts of the coupling can cause interference between the bore and shaft, preventing proper operation of the coupling.

Coupling Installation

Once the inspection shows the coupling parts to be free of defects, the hubs can be mounted on their respective shafts. If it is necessary to heat the hubs to achieve the proper interference fit, an oil or water bath should be used. Spot heating with a flame or torch should be avoided because it causes distortion and may adversely affect the hubs.

Care must be exercised during installation of a new coupling or the reassembly of an existing unit. Keys and keyways should be coated with a sealing compound that is resistant to the lubricant used in the coupling. Seals should be inspected to ensure that they are pliable and in good condition. They must be installed properly in the sleeve with the lip in good contact with the hub. Sleeve flange gaskets must be whole, in good condition, clean, and free of nicks or cracks. Lubrication plugs must be cleaned before being installed and must fit tightly.

The specific installation procedure is dependent on the type and mounting configuration of the coupling. However, common elements of all coupling installations include: spacing, bolting, lubrication, and the use of matching parts. The sections to follow discuss these installation elements.

Spacing

Spacing between the mating parts of the coupling must be within manufacturer's tolerances. For example, an elastomeric coupling must have a specific distance between the coupling faces. This distance determines the position of the rubber boot that provides transmission of power from the driver to the driven machine component. If this distance is not exact, the elastomer will attempt to return to its relaxed position, inducing excessive axial movement in both shafts.

Bolting

Couplings are designed to use a specific type of bolt. Coupling bolts have a hardened cylindrical body sized to match the assembled coupling width. Hardened bolts are required because standard bolts do not have the tensile strength to absorb the torsional and shearing loads in coupling applications and may fail, resulting in coupling failure and machine-train damage.

Lubrication

Most couplings require lubrication, and care must be taken to ensure that the proper type and quantity is used during the installation process. Inadequate or improper lubrication reduces coupling reliability and reduces its useful life. In addition, improper lubrication can cause serious damage to the machine-train. For example, when a gear-type coupling is over filled with grease, the coupling will lock. In most cases, its locked position will increase the vibration level and induce an abnormal loading on the bearings of both the driver and driven unit, resulting in bearing failure.

Matching Parts

Couplings are designed for a specific range of applications, and proper perform-ance depends on the total design of the coupling system. As a result, it is generally not a good practice to mix coupling types. Note, however, that it is common practice in some steel industry applications to use coupling halves from two different types of couplings. For example, a rigid coupling half is sometimes mated to a flexible coupling half, creating a hybrid. While this approach may provide short-term power transmission, it can result in an increase in the number, frequency, and severity of machine-train problems.

Coupling Alignment

The last step in the installation process is verifying coupling and shaft alignment. With the exception of special application couplings such as spindles and jackshafts, all couplings must be aligned within relatively close tolerances (i.e., 0.001–0.002 in.).

LUBRICATION AND MAINTENANCE

Couplings require regular lubrication and maintenance to ensure optimum trouble-free service life. When proper maintenance is not conducted, premature coupling failure and/or damage to machine-train components such as bearings can be expected.

Determining Cause of Failure

When a coupling failure occurs, it is important to determine the cause of failure. Failure may result from a coupling defect, an external condition, or workman-ship during installation.

Most faults are attributed to poorly machined surfaces causing out-of-specification tolerances, although defective material failures also occur. Inadequate material hardness and poor strength factors contribute to many pre-mature failures. Other common causes are improper coupling selection, improper installation, and/or excessive misalignment.

Lubrication Requirements

Lubrication requirements vary depending on application and coupling type. Because rigid couplings do not require lubrication, this section discusses lubrica-tion requirements for mechanical-flexing, material-flexing, and combination flexible couplings only.

Mechanical-Flexing Couplings

It is important to follow the manufacturer's instructions for lubricating mechanical-flexing couplings, which must be lubricated internally. Lubricant seals must be in good condition and properly fitted into place. Coupling covers contain the lubricant and prevent contaminants from entering the coupling interior. The covers are designed in two configurations, split either horizontally or vertically. Holes are provided in the covers to allow lubricant to be added without coupling disassembly.

Gear couplings are one type of mechanical-flexing coupling, and there are several ways to lubricate them: grease pack, oil fill, oil collect, and continuous oil flow. Either grease or oil can be used at speeds of 3,600 rpm to 6,000 rpm. Oil is normally used as the lubricant in couplings operating over 6,000 rpm. Grease and oil-lubricated units have end gaskets and seals, which are used to contain the lubricant and seal out the entry of contaminants. The sleeves have lubrication holes, which permit flushing and re-lubrication without disturbing the sleeve gasket or seals.

Material-Flexing Couplings

Material-flexing couplings are designed to be lubrication free.

Combination Couplings

Combination (metallic-grid) couplings are lubricated in the same manner as mechanical-flexing couplings.

Periodic Inspections

It is important to perform periodic inspections of all mechanical equipment and systems that incorporate rotating parts, including couplings and clutches.

Mechanical-Flexing Couplings

To maintain coupling reliability, mechanical-flexing couplings require periodic inspections on a time- or condition-based frequency established by the history of the equipment's coupling life or a schedule established by the predictive maintenance engineer. Items to be included in an inspection are listed below. If any of these items or conditions is discovered, the coupling should be evaluated to determine its remaining operational life or be repaired/replaced.

- Inspect lubricant for traces of metal (indicating component wear).
- Visually inspect coupling mechanical components (roller chains and gear teeth, and grid members) for wear and/or fatigue.

- Inspect seals to ensure they are pliable and in good condition. They must be installed properly in the sleeve with the lip in good contact with the hub.
- Sleeve flange gaskets must be whole, in good condition, clean, and free of nicks or cracks.
- Lubrication plugs must be clean (to prevent the introduction of contaminants to the lubricant and machine surfaces) before being installed and must be torqued to the manufacturer's specifications.
- Setscrews and retainers must be in place and tightened to manufacturer's specifications.
- Inspect shaft hubs, keyways, and keys for cracks, breaks, and physical damage.
- Under operating conditions, perform thermographic scans to determine temperature differences on the coupling (indicates misalignment and/or uneven mechanical forces).

Material-Flexing Couplings

Although designed to be lubrication-free, material-flexing couplings also require periodic inspection and maintenance. This is necessary to ensure that the coupling components are within acceptable specification limits. Periodic inspections for the following conditions are required to maintain coupling reliability. If any of these conditions are found, the coupling should be evaluated to determine its remaining operational life or be repaired/replaced.

- Inspect flexing element for signs of wear or fatigue (cracks, element dust or particles).
- Setscrews and retainers must be in place and tightened to manufacturer's specifications.
- Inspect shaft hubs, keyways, and keys for cracks, breaks, and physical damage.
- Under operating conditions, perform thermographic scans for temperature differences on the coupling, which indicates misalignment and/or uneven mechanical forces.

Combination Couplings

Mechanical components (e.g., grid members) should be visually inspected for wear and/or fatigue. In addition to the items for mechanical-flexing couplings presented in Chapter 2, the grid members on metallic-grid couplings should be replaced if any signs of wear are observed.

Rigid Couplings

The mechanical components of rigid couplings (e.g., hubs, bolts, compression sleeves and halves, keyways, and keys) should be visually inspected for cracks, breaks, physical damage, wear, and/or fatigue. Any component having any of these conditions should be replaced.

KEYS, KEYWAYS, AND KEY SEATS

A key is a piece of material, usually metal, placed in machined slots or grooves cut into two axially oriented parts to mechanically lock them together. For example, keys are used in making the coupling connection between the shaft of a driver and a hub or flange on that shaft. Any rotating element whose shaft incorporates such a keyed connection is referred to as a *keyed-shaft rotor*. Keys provide a positive means for transmitting torque between the shaft and coupling hub when a key is properly fitted in the axial groove.

The groove into which a key is fitted is referred to as a *key seat* when referring to shafts and a *keyway* when referring to hubs. Key seating is the actual machine operation of producing key seats. Keyways are normally made on a key eater or by a broach. Key seats are normally made with a rotary or end mill cutter.

Figure 10.15 is an example of a keyed shaft that shows the key size versus the shaft diameter. Because of standardization and interchangeability, keys are generally proportioned with relation to shaft diameter instead of torsional load.

The effective key length, "L" is that portion of the key having full bearing on hub and shaft. Note that the curved portion of the key seat made with a rotary cutter does not provide full key bearing, so "L" does not include this distance. The use of an end mill cutter results in a square-ended key seat.

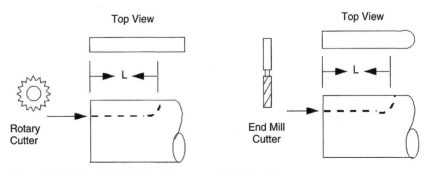

Figure 10.15 Keyed shaft. Key size versus shaft diameter.

Figure 10.16 shows various key shapes: square ends, one square end and one round end, rounded ends, plain taper, and gibe head taper. The majority of keys are square in cross-section, which are preferred through $4\text{-}\frac{1}{2}$ -in. diameter shafts. For bores over $4\text{-}\frac{1}{2}$ in. and thin wall section of hubs, the rectangular (flat) key is used.

The ends are either square, rounded or gibe-head. The gibe-head is usually used with taper keys. If special considerations dictate the use of a keyway in the hub shallower than the preferred square key, it is recommended that the standard rectangular (flat) key be used.

Hub bores are usually straight, although for some special applications, taper bores are sometimes specified. For smaller diameters, bores are designed for clearance fits, and a setscrew is used over the key. The major advantage of a clearance fit is that hubs can be easily assembled and disassembled. For larger diameters, the bores are designed for interference fits without setscrews. For rapid-reversing applications, interference fits are required.

The sections to follow discuss determining keyway depth and width, keyway manufacturing tolerances, key stress calculations, and shaft stress calculations.

DETERMINING KEYWAY DEPTH AND WIDTH

The formula given below and Figure 10.17, Table 10.1 (square keys), and Table 10.2 (flat keys) illustrate how the depth and width of standard square and flat keys and keyways for shafts and hubs are determined.

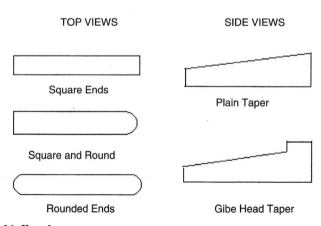

TOP VIEWS   SIDE VIEWS

Square Ends

Plain Taper

Square and Round

Rounded Ends   Gibe Head Taper

Figure 10.16 Key shapes.

Shafts   Hubs

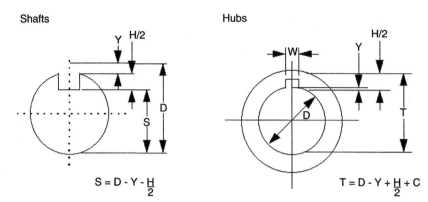

$$S = D - Y - \frac{H}{2}$$

$$T = D - Y + \frac{H}{2} + C$$

Figure 10.17 Shaft and hub dimensions.

Table 10.2 Standard Square Keys and Keyways (inches)*

Diameter Of Holes (Inclusive)	Keyways		Key Stock
	Width	Depth	
$^5/_{16}$ to $^7/_{16}$	$^3/_{32}$	$^3/_{64}$	$^3/_{32} \times {}^3/_{32}$
$^1/_2$ to $^9/_{16}$	$^1/_8$	$^1/_{16}$	$^1/_8 \times {}^1/_8$
$^5/_8$ to $^7/_8$	$^3/_{16}$	$^3/_{32}$	$^3/_{16} \times {}^3/_{16}$
$^{15}/_{16}$ to $1\text{-}^1/_4$	$^1/_4$	$^1/_8$	$^1/_4 \times {}^1/_4$
$1\text{-}^5/_{16}$ to $1\text{-}^3/_8$	$^5/_{16}$	$^5/_{32}$	$^5/_{16} \times {}^5/_{16}$
$1\text{-}^7/_{16}$ to $1\text{-}^3/_4$	$^3/_8$	$^3/_{16}$	$^3/_8 \times {}^3/_8$
$1\text{-}^{13}/_{16}$ to $2\text{-}^1/_4$	$^1/_2$	$^1/_4$	$^1/_2 \times {}^1/_2$
$2\text{-}^5/_{16}$ to $2\text{-}^3/_4$	$^5/_8$	$^5/_{16}$	$^5/_8 \times {}^5/_8$
$2\text{-}^{13}/_{16}$ to $3\text{-}^1/_4$	$^3/_4$	$^3/_8$	$^3/_4 \times {}^3/_4$
$3\text{-}^5/_{16}$ to $3\text{-}^3/_4$	$^7/_8$	$^7/_{16}$	$^7/_8 \times {}^7/_8$
$3\text{-}^{13}/_{16}$ to $4\text{-}^1/_2$	1	$^1/_2$	1×1

*Square keys are normally used through shaft diameter $4\text{-}^1/_2$ in.; larger shafts normally use flat keys.
Source: The Falk Corporation.

Table 10.3 Standard Flat Keys and Keyways (inches)

Diameter Of Holes (Inclusive)	Keyways		Key Stock
	Width	**Depth**	
$\frac{1}{2}$ to $\frac{9}{16}$ "	$\frac{1}{8}$	$\frac{3}{64}$	$\frac{1}{8} \times \frac{1}{32}$
$\frac{5}{8}$ to $\frac{7}{8}$ "	$\frac{3}{16}$	$\frac{1}{16}$	$\frac{3}{16} \times \frac{1}{8}$
$\frac{15}{16}$ to 1-$\frac{1}{4}$ "	$\frac{1}{4}$	$\frac{3}{32}$	$\frac{1}{4} \times \frac{3}{16}$
1-$\frac{5}{16}$ to 1-$\frac{3}{8}$ "	$\frac{5}{16}$	$\frac{1}{8}$	$\frac{5}{16} \times \frac{1}{4}$
1-$\frac{7}{16}$ to 1-$\frac{3}{4}$ "	$\frac{3}{8}$	$\frac{1}{8}$	$\frac{3}{8} \times \frac{1}{4}$
1-$\frac{13}{16}$ to 2-$\frac{1}{4}$ "	$\frac{1}{2}$	$\frac{3}{16}$	$\frac{1}{2} \times \frac{3}{8}$
2-$\frac{5}{16}$ to 2-$\frac{3}{4}$ "	$\frac{5}{8}$	$\frac{7}{32}$	$\frac{5}{8} \times \frac{7}{16}$
2-$\frac{13}{16}$ to 3-$\frac{1}{4}$ "	$\frac{3}{4}$	$\frac{1}{4}$	$\frac{3}{4} \times \frac{1}{2}$
3-$\frac{5}{16}$ to 3-$\frac{3}{4}$ "	$\frac{7}{8}$	$\frac{5}{16}$	$\frac{7}{8} \times \frac{5}{8}$
3-$\frac{13}{16}$ to 4-$\frac{1}{2}$ "	1	$\frac{3}{8}$	$1 \times \frac{3}{4}$
4-$\frac{9}{16}$ to 5-$\frac{1}{2}$ "	1-$\frac{1}{4}$	$\frac{7}{16}$	$1\frac{1}{4} \times \frac{7}{8}$
5-$\frac{9}{16}$ to 6-$\frac{1}{2}$ "	1-$\frac{1}{2}$	$\frac{1}{2}$	1-$\frac{1}{2} \times 1$
6-$\frac{9}{16}$ to 7-$\frac{1}{2}$ "	1-$\frac{3}{4}$	$\frac{5}{8}$	1-$\frac{3}{4} \times \frac{1}{4}$
7-$\frac{9}{16}$ to 9"	2	$\frac{3}{4}$	$2 \times$ 1-$\frac{3}{4}$
9-$\frac{1}{16}$ to 11"	2-$\frac{1}{2}$	$\frac{7}{8}$	2-$\frac{1}{2} \times$ 1-$\frac{3}{4}$
11-$\frac{1}{16}$ to 13"	3	1	3×2
13-$\frac{1}{16}$ to 15"	3-$\frac{1}{2}$	1-$\frac{1}{4}$	3-$\frac{1}{2} \times$ 2-$\frac{1}{2}$
15-$\frac{1}{6}$ to 18"	4	1-$\frac{1}{2}$	4×3
18-$\frac{1}{16}$ to 22"	5	1-$\frac{3}{4}$	$5 \times 3\frac{1}{2}$
22-$\frac{1}{16}$ to 26"	6	4	
26-$\frac{1}{16}$ to 30"	7	5	

Source: The Falk Corporation.

$$Y = \frac{D - \sqrt{D^2 - W^2}}{2}$$

where:

C = Allowance or clearance for key, inches
D = Nominal shaft or bore diameter, inches

H = Nominal key height, inches
W = Nominal key width, inches
Y = Chordal height, inches

Note: Tables shown below are prepared for manufacturing use. Dimensions given are for standard shafts and keyways.

KEYWAY MANUFACTURING TOLERANCES

Keyway manufacturing tolerances (illustrated in Figure 10.18) are referred to as *offset* (centrality) and *lead* (cross axis). Offset or centrality is referred to as Dimension "N"; lead or cross axis is referred to as Dimension "J." Both must be kept within permissible tolerances, usually 0.002 in.

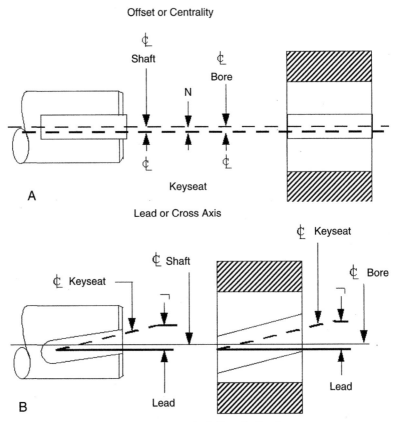

Figure 10.18 Manufacturing tolerances. A, Offset. B, Lead.

KEY STRESS CALCULATIONS

Calculations for shear and compressive key stresses are based on the following assumptions:

1. The force acts at the radius of the shaft.
2. The force is uniformly distributed along the key length.
3. None of the tangential load is carried by the frictional fit between shaft and bore.

The shear and compressive stresses in a key are calculated using the following equations (see Figure 10.19):

$$Ss = \frac{2T}{(d) \times (w) \times (L)} \qquad Sc = \frac{2T}{(d) \times (h_1) \times (L)}$$

where:

d = Shaft diameter, inches (use average diameter for taper shafts)

h_1 = Height of key in the shaft or hub that bears against the keyway, inches. Should equal h_2 for square keys. For designs where unequal portions of the key are in the hub or shaft, h_1 is the minimum portion.

Hp = Power, horsepower

L = Effective length of key, inches

RPM = Revolutions per minute

Ss = Shear stress, psi

Sc = Compressive stress, psi

T = Shaft torque, lb-in. or $\dfrac{Hp \times 63000}{RPM}$

w = Key width, inches

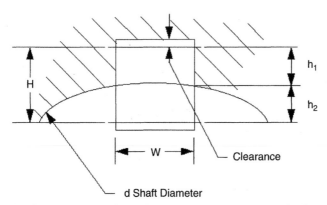

Figure 10.19** **Measurements used in calculating shear and compressive key stress.

Key material is usually AISI 1018 or AISI 1045. Table 10.4 provides the allowable stresses for these materials.

Example: Select a key for the following conditions: 300 Hp at 600 RPM; 3-inch diameter shaft, $\frac{3}{4}$-inch $\times$ $\frac{3}{4}$-inch key, 4-inch key engagement length.

$$T = Torque = \frac{Hp \times 63,000}{RPM} = \frac{300 \times 63,000}{600} = 31,500 \text{ in-lbs}$$

$$Ss = \frac{2T}{d \times w \times L} = \frac{2 \times 31,500}{3 \times 3/4 \times 4} = 7,000 \text{ psi}$$

$$Sc = \frac{2T}{d \times h_1 \times L} = \frac{2 \times 31,500}{3 \times 3/8 \times 4} = 14,000 \text{ psi}$$

The AISI 1018 key can be used since it is within allowable stresses listed in Table 10.4 (allowable Ss = 7,500, allowable Sc = 5,000).

Note: If shaft had been 2-$\frac{3}{4}$-in. diameter (4-in. hub), the key would be $\frac{5}{8}$-in. $\times$ $\frac{5}{8}$-in., Ss = 9,200 psi, Sc = 18,400 psi, and a heat-treated key of AISI 1045 would have been required (allowable Ss = 15,000, allowable Sc = 30,000).

SHAFT STRESS CALCULATIONS

Torsional stresses are developed when power is transmitted through shafts. In addition, the tooth loads of gears mounted on shafts create bending stresses. Shaft design, therefore, is based on safe limits of torsion and bending.

To determine minimum shaft diameter in inches:

$$Minimum \ Shaft \ Diameter = \sqrt[3]{\frac{Hp \times 321000}{RPM \times Allowable \ Stress}}$$

Table 10.4 Allowable Stresses for AISI 1018 and AISI 1045

Material	Heat Treatment	Allowable Stresses – psi	
		Shear	Compressive
AISI 1018	None	7,500	15,000
AISI 1045	255-300 Bhn	15,000	30,000

Source: The Falk Corporation.

Example:

Hp = 300
RPM = 30
Material = 225 Brinell

From Figure 10.20 at 225 Brinell, Allowable Torsion = 8000 psi

$$Minimum\ Shaft\ Diameter = \sqrt[3]{\frac{300 \times 321000}{30 \times 8000}} = \sqrt[3]{402} = 7.38\ \text{inches}$$

From Table 10.5, note that the cube of 7-$\frac{1}{4}$ in. is 381, which is too small (i.e., <402) for this example. The cube of 7-$\frac{1}{2}$ in. is 422, which is large enough.

To determine shaft stress, psi:

$$Shaft\ Stress = \frac{Hp \times 321,000}{RPM \times d^3}$$

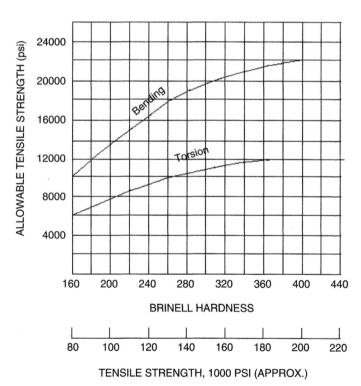

Figure 10.20 Allowable stress as a function of Brinell hardness.

Table 10.5   Shaft Diameters (Inches) and Their Cubes (Cubic Inches)

D	D³	D	D³	D	D³
1	1.00	5	125.0	9	729
1-¼	1.95	5-¼	145	9-½	857
1-½	3.38	5-½	166.4	10	1000
1-¾	5.36	5-¾	190.1	10-½	1157
2	8.00	6	216	11	1331
2-¼	11.39	6-¼	244	11-½	1520
2-½	15.63	6-½	275	12	1728
2-¾	20.80	6-¾	308	12-½	1953
3	27.00	7	343	13	2197
3-¼	34.33	7-¼	381	14	2744
3-½	42.88	7-½	422	15	3375
3-¾	52.73	7-¾	465	16	4096
4	64.00	8	512	17	4913
4-¼	76.77	8-¼	562	18	5832
4-½	91.13	8-½	614	19	6859
4-¾	107.2	8-¾	670	20	8000

Source: The Falk Corporation.

Example: Given 7-$\frac{1}{2}$ -in. shaft for 300 Hp at 30 RPM

$$Shaft\ Stress = \frac{300 \times 321,000}{30 \times (7 - \frac{1}{2})^3} = 7,600\ psi$$

Note: The 7-$\frac{1}{4}$ -in. diameter shaft would be stressed to 8420 psi

11

GEARS AND GEARBOXES

A gear is a form of disc, or wheel, that has teeth around its periphery for the purpose of providing a positive drive by meshing the teeth with similar teeth on another gear or rack.

SPUR GEARS

The spur gear might be called the basic gear since all other types have been developed from it. Its teeth are straight and parallel to the center bore line, as shown in Figure 11.1. Spur gears may run together with other spur gears or parallel shafts, with internal gears on parallel shafts, and with a rack. A rack such as the one illustrated in Figure 11.2 is in effect a straight-line gear. The smallest of a pair of gears (Figure 11.3) is often called a *pinion*.

The involute profile or form is the one most commonly used for gear teeth. It is a curve that is traced by a point on the end of a taut line unwinding from a circle. The larger the circle, the straighter the curvature; for a rack, which is essentially a section of an infinitely large gear, the form is straight or flat. The generation of an involute curve is illustrated in Figure 11.4.

The involute system of spur gearing is based on a rack having straight, or flat, sides. All gears made to run correctly with this rack will run with each other. The sides of each tooth incline toward the center top at an angle called the *pressure angle*, shown in Figure 11.5.

The 14.5-degree pressure angle was standard for many years. In recent years, however, the use of the 20-degree pressure angle has been growing, and today,

Figure 11.1 Example of a spur gear.

Figure 11.2 Rack or straight-line gear.

Figure 11.3 Typical spur gears.

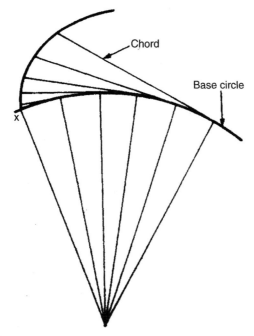

Figure 11.4 Invlute curve.

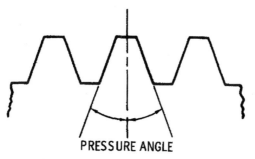

Figure 11.5 Pressure angle.

14.5-degree gearing is generally limited to replacement work. The principal reasons are that a 20-degree pressure angle results in a gear tooth with greater strength and wear resistance and permits the use of pinions with a few fewer teeth. The effect of the pressure angle on the tooth of a rack is shown in Figure 11.6.

It is extremely important that the pressure angle be known when gears are mated, as all gears that run together must have the same pressure angle. The pressure angle of a gear is the angle between the line of action and the line tangent to the

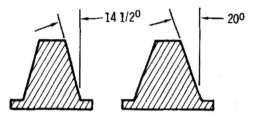

Figure 11.6 Different pressure angles on gear teeth.

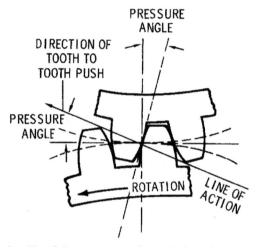

Figure 11.7 Relationship of the pressure angle to the line of action.

pitch circles of mating gears. Figure 11.7 illustrates the relationship of the pressure angle to the line of action and the line tangent to the pitch circles.

PITCH DIAMETER AND CENTER DISTANCE

Pitch circles have been defined as the imaginary circles that are in contact when two standard gears are in correct mesh. The diameters of these circles are the pitch diameters of the gears. The center distance of the two gears, therefore, when correctly meshed, is equal to one half of the sum of the two pitch diameters, as shown in Figure 11.8.

This relationship may also be stated in an equation and may be simplified by using letters to indicate the various values, as follows:

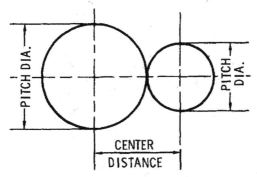

Figure 11.8 Pitch diameter and center distance.

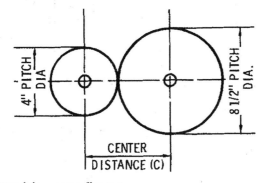

Figure 11.9 Determining center distance.

C = Center distance
D_1 = First pitch diameter
D_2 = Second pitch diameter

$$C = \frac{D_1 + D_2}{2} \qquad D_1 = 2C - D_2 \qquad D_2 = 2C - D_1$$

Example: The center distance can be found if the pitch diameters are known (Figure 11.9).

CIRCULAR PITCH

A specific type of pitch designates the size and proportion of gear teeth. In gearing terms, there are two specific types of pitch: circular pitch and diametrical pitch. Circular pitch is simply the distance from a point on one tooth to a corresponding point on the next tooth, measured along the pitch line or circle, as illustrated in Figure 11.10. Large-diameter gears are frequently made to circular pitch dimensions.

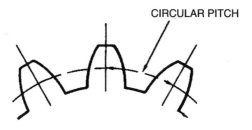

CIRCULAR PITCH

Figure 11.10

DIAMETRICAL PITCH AND MEASUREMENT

The diametrical pitch system is the most widely used, as practically all common-sized gears are made to diametrical pitch dimensions. It designates the size and proportions of gear teeth by specifying the number of teeth in the gear for each inch of the gear's pitch diameter. For each inch of pitch diameter, there are pi (π) inches, or 3.1416 in., of pitch-circle circumference. The diametric pitch number also designates the number of teeth for each 3.1416 in. of pitch-circle circumference. Stated in another way, the diametrical pitch number specifies the number of teeth in 3.1416 in. along the pitch line of a gear.

For simplicity of illustration, a whole-number pitch-diameter gear (4 in.), is shown in Figure 11.11.

Figure 11.11 illustrates that the diametrical pitch number specifying the number of teeth per inch of pitch diameter must also specify the number of

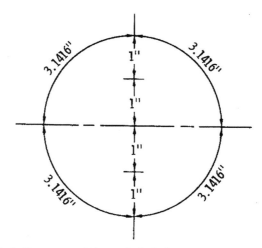

Figure 11.11 Pitch diameter and diametrical pitch.

teeth per 3.1416 in. of pitch-line distance. This may be more easily visualized and specifically dimensioned when applied to the rack in Figure 11.12.

Because the pitch line of a rack is a straight line, a measurement can be easily made along it. In Figure 11.12, it is clearly shown that there are 10 teeth in 3.1416 in.; therefore the rack illustrated is a 10 diametrical pitch rack.

A similar measurement is illustrated in Figure 11.13, along the pitch line of a gear. The diametrical pitch being the number of teeth in 3.1416 in. of pitch line, the gear in this illustration is also a 10 diametrical pitch gear.

In many cases, particularly in machine repair work, it may be desirable for the mechanic to determine the diametrical pitch of a gear. This may be done very easily without the use of precision measuring tools, templates, or gauges. Measurements need not be exact because diametrical pitch numbers are usually whole numbers. Therefore, if an approximate calculation results in a value close to a whole number, that whole number is the diametrical pitch number of the gear.

The following three methods may be used to determine the approximate diametrical pitch of a gear. A common steel rule, preferably flexible, is adequate to make the required measurements.

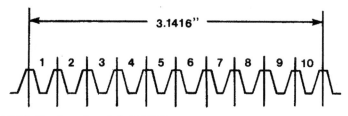

Figure 11.12 Number of teeth in 3.1416 in.

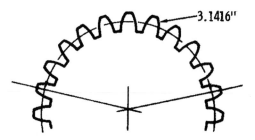

Figure 11.13 Number of teeth in 3.1416 in. on the pitch circle.

METHOD 1

Count the number of teeth in the gear, add 2 to this number, and divide by the outside diameter of the gear. Scale measurement of the gear to the closest fractional size is adequate accuracy.

Figure 11.14 illustrates a gear with 56 teeth and an outside measurement of $5/13$ 16 in. Adding 2 to 56 gives 58; dividing 58 by 5-$^{13}/_{16}$ gives an answer of 9-$^{31}/_{32}$. Since this is approximately 10, it can be safely stated that the gear is a 10 decimal pitch gear.

METHOD 2

Count the number of teeth in the gear and divide this number by the measured pitch diameter. The pitch diameter of the gear is measured from the root or bottom of a tooth space to the top of a tooth on the opposite side of the gear.

Figure 11.15 illustrates a gear with 56 teeth. The pitch diameter measured from the bottom of the tooth space to the top of the opposite tooth is 5-$^{5}/_{8}$ in. Dividing 56 by 5-$^{5}/_{8}$ gives an answer of 9-$^{15}/_{16}$ in. or approximately 10. This method also indicates that the gear is a 10 decimal pitch gear.

PITCH CALCULATIONS

Diametrical pitch, usually a whole number, denotes the ratio of the number of teeth to a gear's pitch diameter. Stated another way, it specifies the number of teeth in a gear for each inch of pitch diameter. The relationship of pitch

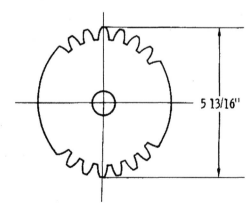

Figure 11.14 Use of Method 1 to approximate the diametrical pitch. In this method the outside diameter of the gear is measured.

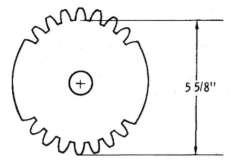

5 5/8"

Figure 11.15 Use of Method 2 to approximate the diametrical pitch. This method uses the pitch diameter of the gear.

diameter, diametrical pitch, and number of teeth can be stated mathematically as follows.

$$P = \frac{N}{D} \qquad D = \frac{N}{P} \qquad N = D \times P$$

where,

D = Pitch diameter
P = Diametrical pitch
N = Number of teeth

If any two values are known, the third may be found by substituting the known values in the appropriate equation.

Example 1: What is the diametrical pitch of a 40-tooth gear with a 5-in. pitch diameter?

$$P = \frac{N}{D} \text{ or } P = \frac{40}{5} \text{ or } P = 8 \text{ diametrical pitch}$$

Example 2: What is the pitch diameter of a 12 diametrical pitch gear with 36 teeth?

$$D = \frac{N}{P} \text{ or } D = \frac{36}{12} \text{ or } D = 3\text{-in. pitch diameter}$$

Example 3: How many teeth are there in a 16 diametrical pitch gear with a pitch diameter of $3\text{-}\frac{3}{4}$ in.?

$$N = D \times P \text{ or } N = 3 - 3/4 \times 16 \text{ or } N = 60 \text{ teeth}$$

Circular pitch is the distance from a point on a gear tooth to the corresponding point on the next gear tooth measured along the pitch line. Its value is equal to the circumference of the pitch circle divided by the number of teeth in the gear. The relationship of the circular pitch to the pitch-circle circumference, number of teeth, and the pitch diameter may also be stated mathematically as follows:

$$\text{Circumference of pitch circle} = \pi D$$

$$P = \frac{\pi D}{N} \qquad D = \frac{PN}{\pi} \qquad N = \frac{\pi D}{P}$$

where,

D = Pitch diameter

N = Number of teeth

P = Circular pitch

π = pi, or 3.1416

If any two values are known, the third may be found by substituting the known values in the appropriate equation.

Example 1: What is the circular pitch of a gear with 48 teeth and a pitch diameter of 6 in.?

$$P = \frac{\pi D}{N} \text{ or } \frac{3.1416 \times 6}{48} \text{ or } \frac{3.1416}{8} \text{ or } P = .3927 \text{ inches}$$

Example 2: What is the pitch diameter of a 0.500-in. circular-pitch gear with 128 teeth?

$$D = \frac{PN}{\pi} \text{ or } \frac{.5 \times 128}{3.1416} \qquad D = 20.371 \text{ inches}$$

The list that follows contains just a few names of the various parts given to gears. These parts are shown in Figures 11.16 and 11.17.

- Addendum: Distance the tooth projects above, or outside, the pitch line or circle.
- Dedendum: Depth of a tooth space below, or inside, the pitch line or circle.

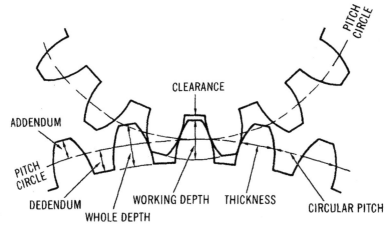

Figure 11.16 Names of gear parts.

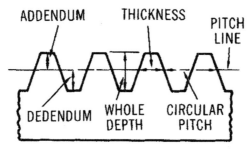

Figure 11.17 Names of rack parts.

- Clearance: Amount by which the dedendum of a gear tooth exceeds the addendum of a matching gear tooth.
- Whole Depth: The total height of a tooth or the total depth of a tooth space.
- Working Depth: The depth of tooth engagement of two matching gears. It is the sum of their addendums.
- Tooth Thickness: The distance along the pitch line or circle from one side of a gear tooth to the other.

Tooth Proportions

The full-depth involute system is the gear system in most common use. The formulas (with symbols) shown below are used for calculating tooth proportions of full-depth involute gears. Diametrical pitch is given the symbol P as before.

Addendum, $a = \dfrac{1}{P}$

Whole Depth, $W_d = \dfrac{2.0 + .002}{P}$ (20P or smaller)

Dedendum, $W_d = \dfrac{2.157}{P}$ (Larger than 20P)

Whole Depth, $b = Wd - a$

Clearance, $c = b - a$

Tooth Thickness, $t = \dfrac{1.5708}{P}$

BACKLASH

Backlash in gears is the play between teeth that prevents binding. In terms of tooth dimensions, it is the amount by which the width of tooth spaces exceeds the thickness of the mating gear teeth. Backlash may also be described as the distance, measured along the pitch line, that a gear will move when engaged with another gear that is fixed or immovable, as illustrated in Figure 11.18.

Normally there must be some backlash present in gear drives to provide running clearance. This is necessary because binding of mating gears can result in heat generation, noise, abnormal wear, possible overload, and/or failure of the drive. A small amount of backlash is also desirable because of the dimensional variations involved in practical manufacturing tolerances.

Backlash is built into standard gears during manufacture by cutting the gear teeth thinner than normal by an amount equal to one half the required figure. When two gears made in this manner are run together, at standard center distance, their allowances combine, provided the full amount of backlash is required.

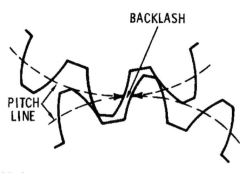

Figure 11.18 Backlash.

On non-reversing drives or drives with continuous load in one direction, the increase in backlash that results from tooth wear does not adversely affect operation. However, on reversing drive and drives where timing is critical, excessive backlash usually cannot be tolerated.

OTHER GEAR TYPES

Many styles and designs of gears have been developed from the spur gear. While they are all commonly used in industry, many are complex in design and manufacture. Only a general description and explanation of principles will be given, as the field of specialized gearing is beyond the scope of this book. Commonly used styles will be discussed sufficiently to provide the millwright or mechanic with the basic information necessary to perform installation and maintenance work.

BEVEL AND MITER

Two major differences between bevel gears and spur gears are their shape and the relation of the shafts on which they are mounted. The shape of a spur gear is essentially a cylinder, while the shape of a bevel gear is a cone. Spur gears are used to transmit motion between parallel shafts, while bevel gears transmit motion between angular or intersecting shafts. The diagram in Figure 11.19 illustrates the bevel gear's basic cone shape. Figure 11.20 shows a typical pair of bevel gears.

Special bevel gears can be manufactured to operate at any desired shaft angle, as shown in Figure 11.21. Miter gears are bevel gears with the same number of teeth in both gears operating on shafts at right angles or at 90 degrees, as shown in Figure 11.22.

A typical pair of straight miter gears is shown in Figure 11.23. Another style of miter gears having spiral rather than straight teeth is shown in Figure 11.24. The spiral-tooth style will be discussed later.

The diametrical pitch number as is done with spur gears establishes the tooth size of bevel gears. Because the tooth size varies along its length, it must be measured at a given point. This point is the outside part of the gear where the tooth is the largest. Because each gear in a set of bevel gears must have the same angles and tooth lengths, as well as the same diametrical pitch, they are manufactured and distributed only in mating pairs. Bevel gears, like spur gears, are manufactured in both the 14.5-degree and 20-degree pressure-angle designs.

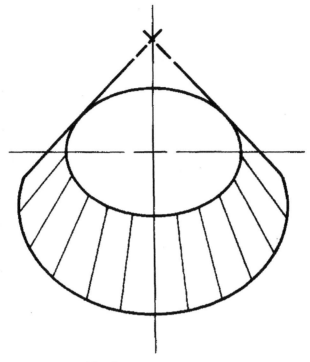

Figure 11.19 Basic shape of bevel gears.

Figure 11.20 Typical set of bevel gears.

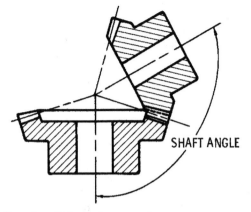

Figure 11.21 Shaft angle, which can be at any degree.

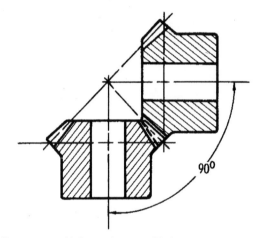

Figure 11.22 Miter gears, which are shown at 90 degrees.

HELICAL

Helical gears are designed for parallel-shaft operation like the pair in Figure 11.25. They are similar to spur gears except that the teeth are cut at an angle to the centerline. The principal advantage of this design is the quiet, smooth action that results from the sliding contact of the meshing teeth. A disadvantage, however, is the higher friction and wear that accompanies this sliding action. The angle at which the gear teeth are cut is called the *helix angle* and is illustrated in Figure 11.26.

It is very important to note that the helix angle may be on either side of the gear's centerline. Or if compared with the helix angle of a thread, it may be either a "right-hand" or a "left-hand" helix. The hand of the helix is the same regardless

Figure 11.23 Typical set of miter gears.

of how viewed. Figure 11.27 illustrates a helical gear as viewed from opposite sides; changing the position of the gear cannot change the hand of the tooth's helix angle. A pair of helical gears, as illustrated in Figure 11.25, must have the same pitch and helix angle but must be of opposite hands (one right hand and one left hand).

Helical gears may also be used to connect nonparallel shafts. When used for this purpose, they are often called "spiral" gears or crossed-axis helical gears. This style of helical gearing is shown in Figure 11.28.

WORM

The worm and worm gear, illustrated in Figure 11.29, are used to transmit motion and power when a high-ratio speed reduction is required. They provide a steady quiet transmission of power between shafts at right angles. The worm is

Figure 11.24 Miter gears with spiral teeth.

Figure 11.25 Typical set of helical gears.

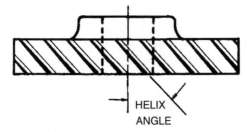

Figure 11.26 The angle at which teeth are cut.

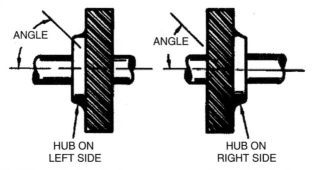

Figure 11.27 Helix angle of teeth: the same no matter from which side the gear is viewed.

Figure 11.28 Typical set of spiral gears.

Figure 11.29 Typical set of worm gears.

always the driver and the worm gear the driven member. Like helical gears, worms and worm gears have "hand." The hand is determined by the direction of the angle of the teeth. Thus, for a worm and worm gear to mesh correctly, they must be the same hand.

The most commonly used worms have either one, two, three, or four separate threads and are called single, double, triple, and quadruple thread worms. The number of threads in a worm is determined by counting the number of starts or entrances at the end of the worm. The thread of the worm is an important feature in worm design, as it is a major factor in worm ratios. The ratio of a mating worm and worm gear is found by dividing the number of teeth in the worm gear by the number of threads in the worm.

HERRINGBONE

To overcome the disadvantage of the high end thrust present in helical gears, the herringbone gear, illustrated in Figure 11.30, was developed. It consists simply of two sets of gear teeth, one right hand and one left hand, on the same gear. The gear teeth of both hands cause the thrust of one set to cancel out the thrust of

Figure 11.30 Herringbone gear.

the other. Thus the advantage of helical gears is obtained, and quiet, smooth operation at higher speeds is possible. Obviously they can only be used for transmitting power between parallel shafts.

GEAR DYNAMICS AND FAILURE MODES

Many machine-trains utilize gear drive assemblies to connect the driver to the primary machine. Gears and gearboxes typically have several vibration spectra associated with normal operation. Characterization of a gearbox's vibration signature box is difficult to acquire but is an invaluable tool for diagnosing machine-train problems. The difficulty is that (1) it is often difficult to mount the transducer close to the individual gears, and (2) the number of vibration sources in a multi-gear drive results in a complex assortment of gear mesh, modulation, and running frequencies. Severe drive-train vibrations (gearbox) are usually due to resonance between a system's natural frequency and the speed of some shaft. The resonant excitation arises from, and is proportional to, gear inaccuracies that cause small periodic fluctuations in pitch-line velocity. Complex machines usually have many resonance zones within their operating speed range because each shaft can excite a system resonance. At resonance these cyclic excitations may cause large vibration amplitudes and stresses.

Basically, forcing torque arising from gear inaccuracies is small. However, under resonant conditions torsional amplitude growth is restrained only by damping in that mode of vibration. In typical gearboxes this damping is often small and permits the gear-excited torque to generate large vibration amplitudes under resonant conditions.

One other important fact about gear sets is that all gear sets have a designed preload and create an induced load (thrust) in normal operation. The direction, radial or axial, of the thrust load of typical gear sets will provide some insight into the normal preload and induced loads associated with each type of gear.

To implement a predictive maintenance program, a great deal of time should be spent understanding the dynamics of gear/gearbox operation and the frequencies typically associated with the gearbox. As a minimum, the following should be identified.

Gears generate a unique dynamic profile that can be used to evaluate gear condition. In addition, this profile can be used as a tool to evaluate the operating dynamics of the gearbox and its related process system.

Gear Damage

All gear sets create a frequency component, called *gear mesh*. The fundamental gear mesh frequency is equal to the number of gear teeth times the running speed of the shaft. In addition, all gear sets will create a series of side bands or modulations that will be visible on both sides of the primary gear mesh frequency. In a normal gear set, each of the side bands will be spaced at exactly the 1X or running speed of the shaft and the profile of the entire gear mesh will be symmetrical.

Normal Profile

In a normal gear set, each of the side bands will be spaced at exactly the 1X running speed of the input shaft, and the entire gear mesh will be symmetrical. In addition, the side bands will always occur in pairs, one below and one above the gear mesh frequency. The amplitude of each of these pairs will be identical. For example, the side band pair indicated as −1 and +1 in Figure 11.31 will be spaced at exactly input speed and have the same amplitude.

If the gear mesh profile were split by drawing a vertical line through the actual mesh (i.e., number of teeth times the input shaft speed), the two halves would be exactly identical. Any deviation from a symmetrical gear mesh profile is indicative of a gear problem. However, care must be exercised to ensure that the problem is internal to the gears and induced by outside influences. External misalignment, abnormal induced loads, and a variety of other outside influences will destroy the symmetry of the gear mesh profile. For example, the single reduction gearbox used to transmit power to the mold oscillator system on a continuous caster drives two eccentrics. The eccentric rotation of these two cams is transmitted directly into the gearbox and will create the appearance of

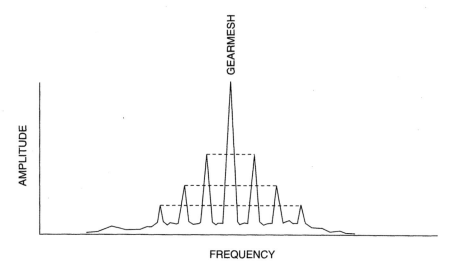

Figure 11.31 Normal profile is symmetrical.

eccentric meshing of the gears. The spacing and amplitude of the gear mesh profile will be destroyed by this abnormal induced load.

Excessive Wear

Figure 11.32 illustrates a typical gear profile with worn gears. Note that the spacing between the side bands becomes erratic and they are no longer spaced at the input shaft speed. The side bands will tend to vary between the input and output speeds but will not be evenly spaced.

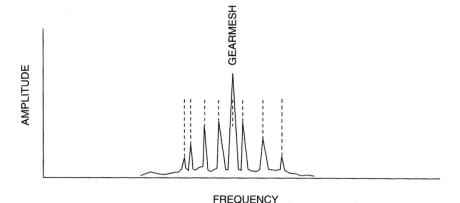

Figure 11.32 Wear or excessive clearance changes side band spacing.

In addition to gear tooth wear, center-to-center distance between shafts will create an erratic spacing and amplitude. If the shafts are too close together, the spacing will tend to be at input shaft speed, but the amplitude will drop drastically. Because the gears are deeply meshed (i.e., below the normal pitch line), the teeth will maintain contact through the entire mesh. This loss of clearance will result in lower amplitudes but will exaggerate any tooth profile defect that may be present.

If the shafts are too far apart, the teeth will mesh above the pitch line. This type of meshing will increase the clearance between teeth and amplify the energy of the actual gear mesh frequency and all of its side bands. In addition, the load-bearing characteristics of the gear teeth will be greatly reduced. Since the pressure is focused on the tip of each tooth, there is less cross-section and strength in the teeth. The potential for tooth failure is increased in direct proportion the amount of excess clearance between shafts.

Cracked Or Broken Tooth

Figure 11.33 illustrates the profile of a gear set with a broken tooth. As the gear rotates, the space left by the chipped or broken tooth will increase the mechanical clearance between the pinion and bull gear. The result will be a low amplitude side band that will occur to the left of the actual gear mesh frequency. When the next, undamaged teeth mesh, the added clearance will result in a higher-energy impact.

The resultant side band, to the right of the mesh frequency, will have much higher amplitude. The paired side bands will have non-symmetrical amplitude that represents this disproportional clearance and impact energy.

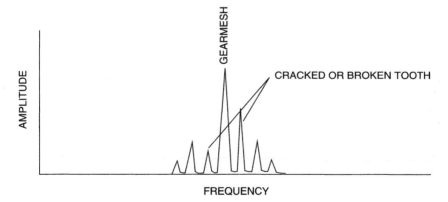

Figure 11.33 A broken tooth will produce an asymmetrical side band profile.

If the gear set develops problems, the amplitude of the gear mesh frequency will increase and the symmetry of the side bands will change. The pattern illustrated In Figure 11.34 is typical of a defective gear set. Note the asymmetrical relationship of the side bands.

COMMON CHARACTERISTICS

You should have a clear understanding of the types of gears generally utilized in today's machinery, how they interact, and the forces they generate on a rotating shaft. There are two basic classifications of gear drives: (1) shaft centers parallel, and (2) shaft centers not parallel. Within these two classifications are several typical gear types.

Shaft Centers Parallel

There are four basic gear types that are typically used in this classification. All are mounted on parallel shafts and, unless an idler gear is also used, will have opposite rotation between the drive and driven gear (if the drive gear has a clockwise rotation, then the driven gear will have a counter-clockwise rotation). The gear sets commonly used in machinery include the following.

Spur Gears

The shafts are in the same plane and parallel. The teeth are cut straight and parallel to the axis of the shaft rotation. No more than two sets of teeth are in

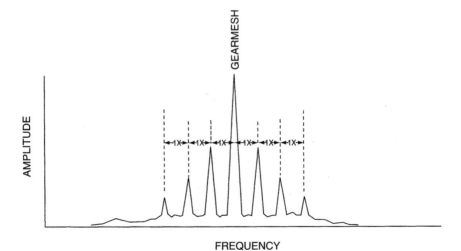

Figure 11.34 Typical defective gear mesh signature.

mesh at one time, so the load is transferred from one tooth to the next tooth rapidly. Usually spur gears are used for moderate to low speed applications. Rotation of spur gear sets is opposite unless one or more idler gears are included in the gearbox. Typically, spur gear sets will generate a radial load (preload) opposite the mesh on their shaft support bearings and little or no axial load.

Backlash is an important factor in proper spur gear installation. A certain amount of backlash must be built into the gear drive allowing for tolerances in concentricity and tooth form. Insufficient backlash will cause early failure because of overloading.

As indicated in Figure 11.11, spur gears by design have a preload opposite the mesh and generate an induced load, or tangential force (TF) in the direction of rotation. This force can be calculated as:

$$TF = \frac{126,000 * HP}{D_p * RPM}$$

In addition, a spur gear will generate a separating force, S_{TF}, that can be calculated as:

$$S_{TF} = TF * \tan \phi$$

where

TF	= Tangential Force
HP	= Input horsepower to pinion or gear
D_p	= Pitch diameter of pinion or gear
RPM	= Speed of pinion or gear
ϕ	= Pinion or gear tooth pressure angle

Helical Gears

The shafts are in the same plane and parallel but the teeth are cut at an angle to the centerline of the shafts. Helical teeth have an increased length of contact, run quieter and have a greater strength and capacity than spur gears. Normally the angle created by a line through the center of the tooth and a line parallel to the shaft axis is 45 degrees. However, other angles may be found in machinery. Helical gears also have a preload by design; the critical force to be considered, however, is the thrust load (axial) generated in normal operation; see Figure 11.12.

$$TF = \frac{126,000 * HP}{D_p * RPM}$$

$$S_{TF} = \frac{TF * \tan \phi}{\cos \lambda}$$

$$T_{TF} = TF * \tan \lambda$$

where

TF = Tangential Force
S_{TF} = Separating Force
T_{TF} = Thrust Force
HP = Input horsepower to pinion or gear
D_p = Pitch diameter of pinion or gear
RPM = Speed of pinion or gear
ϕ = Pinion or gear tooth pressure angle
λ = Pinion or gear helix angle

Herringbone Gears

These are commonly called *double helical* because they have teeth cut with right and left helix angles. They are used for heavy loads at medium to high speeds. They do not have the inherent thrust forces that are present in helical gear sets. Herringbone gears, by design, cancel the axial loads associated with a single helical gear. The typical loads associated with herringbone gear sets are the radial side-load created by gear mesh pressure and a tangential force in the direction of rotation.

Internal Gears

Internal gears can be run only with an external gear of the same type, pitch, and pressure angle. The preload and induced load will depend on the type of gears used. Refer to spur or helical for axial and radial forces.

TROUBLESHOOTING

One of the primary causes of gear failure is the fact that, with few exceptions, gear sets are designed for operation in one direction only. Failure is often caused by inappropriate bi-directional operation of the gearbox or backward installation of the gear set. Unless specifically manufactured for bi-directional operation, the "non-power" side of the gear's teeth is not finished. Therefore, this side is rougher and does not provide the same tolerance as the finished "power" side.

Note that it has become standard practice in some plants to reverse the pinion or bull gear in an effort to extend the gear set's useful life. While this practice permits longer operation times, the torsional power generated by a reversed gear set is not as uniform and consistent as when the gears are properly installed.

Table 11.1 *Common Failure Modes of Gearboxes and Gear Sets*

THE CAUSES	Gear Failures	Variations In Torsional Power	Insufficient Power Output	Overheated Bearings	Short Bearing Life	Overload on Driver	High Vibration	High Noise Levels	Motor Trips
Bent Shaft				•	•	•	•		
Broken or Loose Bolts or Setscrews				•			•		
Damaged Motor						•	•		•
Elliptical Gears		•	•			•	•		
Exceeds Motor's Brake Horsepower Rating			•			•			
Excessive or Too Little Backlash	•	•							
Excessive Torsional Loading	•	•	•	•	•	•			•
Foreign Object In Gearbox	•						•	•	•
Gear Set Not Suitable for Application	•		•			•	•		
Gears Mounted Backward on Shafts			•				•	•	
Incorrect Center-to-Center Distance Between Shafts							•	•	
Incorrect Direction of Rotation			•			•	•		
Lack of or Improper Lubrication	•	•		•	•		•	•	•
Misalignment of Gears or Gearbox	•	•		•	•		•	•	
Overload	•		•	•	•	•			
Process Induced Misalignment	•	•		•	•				
Unstable Foundation			•	•			•	•	
Water or Chemicals in Gearbox	•								
Worn Bearings							•	•	
Worn Coupling							•		

Source: Integrated Systems, Inc.

Gear overload is another leading cause of failure. In some instances, the overload is constant, which is an indication that the gearbox is not suitable for the application. In other cases, the overload is intermittent and only occurs when the speed changes or when specific production demands cause a momentary spike in the torsional load requirement of the gearbox.

Misalignment, both real and induced, is also a primary root cause of gear failure. The only way to ensure that gears are properly aligned is to hard blue the gears immediately following installation. After the gears have run for a short time, their wear pattern should be visually inspected. If the pattern does not conform to vendor's specifications, alignment should be adjusted.

Poor maintenance practices are the primary source of real misalignment problems. Proper alignment of gear sets, especially large ones, is not an easy task. Gearbox manufacturers do not provide an easy, positive means to ensure that shafts are parallel and that the proper center-to-center distance is maintained.

Induced misalignment is also a common problem with gear drives. Most gearboxes are used to drive other system components, such as bridle or process rolls. If misalignment is present in the driven members (either real or process induced), it also will directly affect the gears. The change in load zone caused by the misaligned driven component will induce misalignment in the gear set. The effect is identical to real misalignment within the gearbox or between the gearbox and mated (i.e., driver and driven) components.

Visual inspection of gears provides a positive means to isolate the potential root cause of gear damage or failures. The wear pattern or deformation of gear teeth provides clues as to the most likely forcing function or cause. The following sections discuss the clues that can be obtained from visual inspection.

NORMAL WEAR

Figure 11.35 illustrates a gear that has a normal wear pattern. Note that the entire surface of each tooth is uniformly smooth above and below the pitch line.

ABNORMAL WEAR

Figures 11.36 through 11.39 illustrate common abnormal wear patterns found in gear sets. Each of these wear patterns suggests one or more potential failure modes for the gearbox.

Figure 11.35 Normal wear pattern.

Figure 11.36 Wear pattern caused by abrasives in lubricating oil.

Abrasion

Abrasion creates unique wear patterns on the teeth. The pattern varies, depending on the type of abrasion and its specific forcing function. Figure 11.36 illustrates severe abrasive wear caused by particulates in the lubricating oil. Note the score marks that run from the root to the tip of the gear teeth.

Chemical Attack or Corrosion

Water and other foreign substances in the lubricating oil supply also cause gear degradation and premature failure. Figure 11.37 illustrates a typical wear pattern on gears caused by this failure mode.

Figure 11.37 Pattern caused by corrosive attack on gear teeth.

Figure 11.38 Pitting caused by gear overloading.

Overloading

The wear patterns generated by excessive gear loading vary, but all share similar components. Figure 11.38 illustrates pitting caused by excessive torsional loading. The pits are created by the implosion of lubricating oil. Other wear patterns, such as spalling and burning, can also help to identify specific forcing functions or root causes of gear failure.

12

COMPRESSORS

Compressors are machines that are used to increase the pressure of a gas or vapor. They can be grouped into two major classifications, centrifugal and positive displacement. This section provides a general discussion of these types of compressors.

CENTRIFUGAL

In general, the centrifugal designation is used when the gas flow is radial and the energy transfer is predominantly due to a change in the centrifugal forces acting on the gas. The force utilized by the centrifugal compressor is the same as that utilized by centrifugal pumps.

In a centrifugal compressor, air or gas at atmospheric pressure enters the eye of the impeller. As the impeller rotates, the gas is accelerated by the rotating element within the confined space that is created by the volute of the compressor's casing. The gas is compressed as more gas is forced into the volute by the impeller blades. The pressure of the gas increases as it is pushed through the reduced free space within the volute.

As in centrifugal pumps, there may be several stages to a centrifugal air compressor. In these multi-stage units, a progressively higher pressure is produced by each stage of compression.

Configuration

The actual dynamics of centrifugal compressors are determined by their design. Common designs are overhung or cantilever, centerline, and bull gear.

Overhung or Cantilever

The cantilever design is more susceptible to process instability than centerline centrifugal compressors. Figure 12.1 illustrates a typical cantilever design.

The overhung design of the rotor (i.e., no outboard bearing) increases the potential for radical shaft deflection. Any variation in laminar flow, volume, or load of the inlet or discharge gas forces the shaft to bend or deflect from its true centerline. As a result, the mode shape of the shaft must be monitored closely.

Centerline

Centerline designs, such as horizontal and vertical split-case, are more stable over a wider operating range, but should not be operated in a variable-demand system. Figure 12.2 illustrates the normal airflow pattern through a horizontal split-case compressor. Inlet air enters the first stage of the compressor, where pressure and velocity increases occur. The partially compressed air is routed to the second stage, where the velocity and pressure are increased further. Adding

Figure 12.1 Cantilever centrifugal compressor is susceptible to instability.

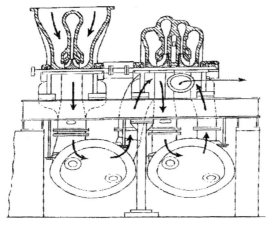

Figure 12.2 Airflow through a centerline centrifugal compressor.

additional stages until the desired final discharge pressure is achieved can continue this process.

Two factors are critical to the operation of these compressors: impeller configuration and laminar flow, which must be maintained through all of the stages.

The impeller configuration has a major impact on stability and operating envelope. There are two impeller configurations, inline and back-to-back, or opposed. With the inline design, all impellers face in the same direction. With the opposed design, impeller direction is reversed in adjacent stages.

Inline A compressor with all impellers facing in the same direction generates substantial axial forces. The axial pressures generated by each impeller for all the stages are additive. As a result, massive axial loads are transmitted to the fixed bearing. Because of this load, most of these compressors use either a Kingsbury thrust bearing or a balancing piston to resist axial thrusting.

Figure 12.3 illustrates a typical balancing piston.

All compressors that use inline impellers must be monitored closely for axial thrusting. If the compressor is subjected to frequent or constant unloading, the axial clearance will increase because of this thrusting cycle. Ultimately, this frequent thrust loading will lead to catastrophic failure of the compressor.

Opposed By reversing the direction of alternating impellers, the axial forces generated by each impeller or stage can be minimized. In effect, the opposed

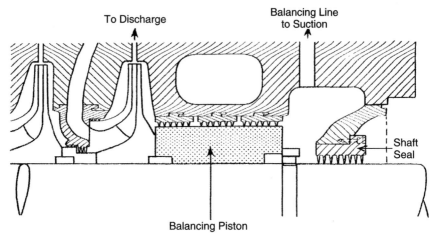

Figure 12.3 Balancing piston resists axial thrust from the inline impeller design of a centerline centrifugal compressor.

impellers tend to cancel the axial forces generated by the preceding stage. This design is more stable and should not generate measurable axial thrusting. This allows these units to contain a normal float and fixed rolling-element bearing.

Bull Gear

The bull gear design uses a direct-driven helical gear to transmit power from the primary driver to a series of pinion-gear-driven impellers that are located around the circumference of the bull gear. Figure 12.4 illustrates a typical bull gear compressor layout.

The pinion shafts are typically a cantilever-type design that has an enclosed impeller on one end and a tilting-pad bearing on the other. The pinion gear is between these two components. The number of impeller-pinions (i.e., stages) varies with the application and the original equipment vendor. However, all bull gear compressors contain multiple pinions that operate in series.

Atmospheric air or gas enters the first-stage pinion, where the pressure is increased by the centrifugal force created by the first-stage impeller. The partially compressed air leaves the first stage, passes through an intercooler, and enters the second-stage impeller. This process is repeated until the fully compressed air leaves through the final pinion-impeller, or stage.

Most bull gear compressors are designed to operate with a gear speed of 3,600 rpm. In a typical four-stage compressor, the pinions operate at progressively higher speeds. A typical range is between 12,000 rpm (first stage) and 70,000 rpm (fourth stage).

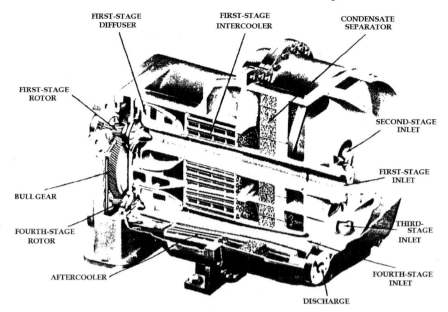

FIRST-STAGE
DIFFUSER

FIRST-STAGE
INTERCOOLER

CONDENSATE
SEPARATOR

FIRST-STAGE
ROTOR

SECOND-STAGE
INLET

FIRST-STAGE
INLET

BULL GEAR

FOURTH-STAGE
ROTOR

THIRD-
STAGE
INLET

AFTERCOOLER

FOURTH-STAGE
INLET

DISCHARGE

Figure 12.4 Bull gear centrifugal compressor.

Because of their cantilever design and pinion rotating speeds, bull gear compressors are extremely sensitive to variations in demand or down-stream pressure changes. Because of this sensitivity, their use should be limited to base load applications.

Bull gear compressors are not designed for, nor will they tolerate, load-following applications. They should not be installed in the same discharge manifold with positive-displacement compressors, especially reciprocating compressors. The standing-wave pulses created by many positive-displacement compressors create enough variation in the discharge manifold to cause potentially serious instability.

In addition, the large helical gear used for the bull gear creates an axial oscillation or thrusting that contributes to instability within the compressor. This axial movement is transmitted throughout the machine-train.

PERFORMANCE

The physical laws of thermodynamics, which define their efficiency and system dynamics, govern compressed-air systems and compressors. This section discusses both the first and second laws of thermodynamics, which apply to all compressors and compressed-air systems. Also applying to these systems are the Ideal Gas Law and the concepts of pressure and compression.

First Law of Thermodynamics

This law states that energy cannot be created or destroyed during a process, such as compression and delivery of air or gas, although it may change from one form of energy to another. In other words, whenever a quantity of one kind of energy disappears, an exactly equivalent total of other kinds of energy must be produced. This is expressed for a steady-flow open system such as a compressor by the following relationship:

Net energy added to system as heat and work $+$ Stored energy of mass entering system $-$ Stored energy of mass leaving system $= 0$

Second Law of Thermodynamics

The second law of thermodynamics states that energy exists at various levels and is available for use only if it can move from a higher to a lower level. For example, it is impossible for any device to operate in a cycle and produce work while exchanging heat only with bodies at a single fixed temperature. In thermodynamics, a measure of the unavailability of energy has been devised and is known as *entropy*. As a measure of unavailability, entropy increases as a system loses heat but remains constant when there is no gain or loss of heat as in an adiabatic process. It is defined by the following differential equation:

$$dS = \frac{dQ}{T}$$

where

$T =$ Temperature (Fahrenheit)
$Q =$ Heat added (BTU)

Pressure/Volume/Temperature (PVT) Relationship

Pressure, temperature, and volume are properties of gases that are completely interrelated. Boyle's Law and Charles's Law may be combined into one equation that is referred to as the Ideal Gas Law. This equation is always true for Ideal gases and is true for real gases under certain conditions.

$$\frac{P_1 V_1}{T_1} = \frac{P_2 V_2}{T_2}$$

For air at room temperature, the error in this equation is less than 1% for pressures as high as 400 psia. For air at one atmosphere of pressure, the error is less than 1% for temperatures as low as $-200°$ Fahrenheit. These error factors will vary for different gases.

Pressure/Compression

In a compressor, pressure is generated by pumping quantities of gas into a tank or other pressure vessel. Progressively increasing the amount of gas in the confined or fixed-volume space increases the pressure. The effects of pressure exerted by a confined gas result from the force acting on the container walls. This force is caused by the rapid and repeated bombardment from the enormous number of molecules that are present in a given quantity of gas.

Compression occurs when the space is decreased between the molecules. Less volume means that each particle has a shorter distance to travel, thus proportionately more collisions occur in a given span of time, resulting in a higher pressure. Air compressors are designed to generate particular pressures to meet specific application requirements.

Other Performance Indicators

The same performance indicators as centrifugal pumps or fans govern centrifugal compressors.

Installation

Dynamic compressors seldom pose serious foundation problems. Since moments and shaking forces are not generated during compressor operation, there are no variable loads to be supported by the foundation. A foundation or mounting of sufficient area and mass to maintain compressor level and alignment and to ensure safe soil loading is all that is required. The units may be supported on structural steel if necessary. The principles defined for centrifugal pumps also apply to centrifugal compressors.

It is necessary to install pressure-relief valves on most dynamic compressors to protect them because of restrictions placed on casing pressure and power input and to keep it out of its surge range. Always install a valve capable of bypassing the full-load capacity of the compressor between its discharge port and the first isolation valve.

Operating Methods

The acceptable operating envelope for centrifugal compressors is very limited. Therefore, care should be taken to minimize any variation in suction supply, backpressure caused by changes in demand, and frequency of unloading. The operating guidelines provided in the compressor vendor's O&M manual should be followed to prevent abnormal operating behavior or premature wear or failure of the system.

Centrifugal compressors are designed to be base loaded and may exhibit abnormal behavior or chronic reliability problems when used in a load-following mode of operation. This is especially true of bull gear and cantilever compressors. For example, a 1-psig change in discharge pressure may be enough to cause catastrophic failure of a bull gear compressor.

Variations in demand or backpressure on a cantilever design can cause the entire rotating element and its shaft to flex. This not only affects the compressor's efficiency but also accelerates wear and may lead to premature shaft or rotor failure.

All compressor types have moving parts, high noise levels, high pressures, and high-temperature cylinder and discharge-piping surfaces.

Positive Displacement

Positive-displacement compressors can be divided into two major classifications, rotary and reciprocating.

Rotary

The rotary compressor is adaptable to direct drive by the use of induction motors or multi-cylinder gasoline or diesel engines. These compressors are compact, relatively inexpensive, and require a minimum of operating attention and maintenance. They occupy a fraction of the space and weight of a reciprocating machine having equivalent capacity.

Configuration

Rotary compressors are classified into three general groups: sliding vane, helical lobe, and liquid-seal ring.

Sliding Vane The basic element of the sliding-vane compressor is the cylindrical housing and the rotor assembly. This compressor, which is illustrated in Figure 12.5, has longitudinal vanes that slide radially in a slotted rotor mounted eccentrically in a cylinder. The centrifugal force carries the sliding vanes against the cylindrical case with the vanes forming a number of individual longitudinal cells in the eccentric annulus between the case and rotor. The suction port is located where the longitudinal cells are largest. The size of each cell is reduced by

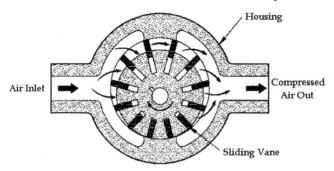

Figure 12.5 Rotary sliding-vane compressor.

the eccentricity of the rotor as the vanes approach the discharge port, thus compressing the gas.

Cyclical opening and closing of the inlet and discharge ports occurs by the rotor's vanes passing over them. The inlet port is normally a wide opening that is designed to admit gas in the pocket between two vanes. The port closes momentarily when the second vane of each air-containing pocket passes over the inlet port.

When running at design pressure, the theoretical operation curves are identical (Figure 12.6) to a reciprocating compressor. However, there is one major difference between a sliding-vane and a reciprocating compressor. The reciprocating unit has spring-loaded valves that open automatically with small pressure differentials between the outside and inside cylinder. The sliding-vane compressor has no valves.

The fundamental design considerations of a sliding-vane compressor are the rotor assembly, cylinder housing, and the lubrication system.

Housing and Rotor Assembly Cast iron is the standard material used to construct the cylindrical housing, but other materials may be used if corrosive conditions exist. The rotor is usually a continuous piece of steel that includes the shaft and is made from bar stock. Special materials can be selected for corrosive applications. Occasionally, the rotor may be a separate iron casting keyed to a shaft. On most standard air compressors, the rotor-shaft seals are semi-metallic packing in a stuffing box. Commercial mechanical rotary seals can be supplied when needed. Cylindrical roller bearings are generally used in these assemblies.

Vanes are usually asbestos or cotton cloth impregnated with a phenolic resin. Bronze or aluminum also may be used for vane construction. Each vane fits into

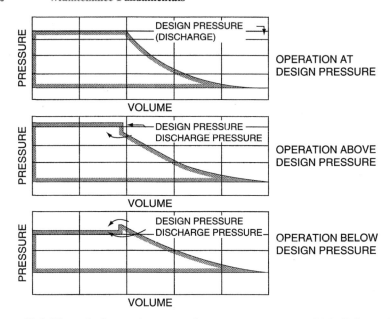

Figure 12.6 Theoretical operation curves for rotary compressors with built-in porting.

a milled slot extending the full length of the rotor and slides radially in and out of this slot once per revolution. Vanes are the most maintenance-prone part in the compressor. There are from 8 to 20 vanes on each rotor, depending on its diameter. A greater number of vanes increases compartmentalization, which reduces the pressure differential across each vane.

Lubrication System A V-belt-driven, force-fed oil lubrication system is used on water-cooled compressors. Oil goes to both bearings and to several points in the cylinder. Ten times as much oil is recommended to lubricate the rotary cylinder as is required for the cylinder of a corresponding reciprocating compressor. The oil carried over with the gas to the line may be reduced 50% with an oil separator on the discharge. Use of an aftercooler ahead of the separator permits removal of 85-90% of the entrained oil.

Helical Lobe or Screw The helical lobe, or screw, compressor is shown in Figure 12.7. It has two or more mating sets of lobe-type rotors mounted in a common housing. The male lobe, or rotor, is usually direct-driven by an electric motor. The female lobe, or mating rotor, is driven by a helical gear set that is mounted on the outboard end of the rotor shafts. The gears provide both motive power for the female rotor and absolute timing between the rotors.

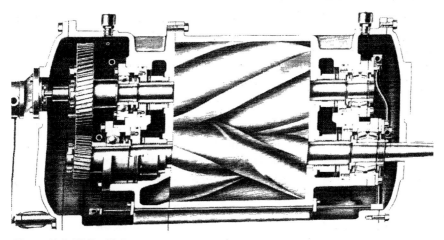

Figure 12.7 Helical lobe, or screw, rotary air compressor.

The rotor set has extremely close mating clearance (i.e., about 0.5 mil) but no metal-to-metal contact. Most of these compressors are designed for oil-free operation. In other words, no oil is used to lubricate or seal the rotors. Instead, oil lubrication is limited to the timing gears and bearings that are outside the air chamber. Because of this, maintaining proper clearance between the two rotors is critical.

This type of compressor is classified as a constant volume, variable-pressure machine that is quite similar to the vane-type rotary in general characteristics. Both have a built-in compression ratio.

Helical-lobe compressors are best suited for base-load applications where they can provide a constant volume and pressure of discharge gas. The only recommended method of volume control is the use of variable-speed motors. With variable-speed drives, capacity variations can be obtained with a proportionate reduction in speed. A 50% speed reduction is the maximum permissible control range.

Helical-lobe compressors are not designed for frequent or constant cycles between load and no-load operation. Each time the compressor unloads, the rotors tend to thrust axially. Even though the rotors have a substantial thrust bearing and, in some cases, a balancing piston to counteract axial thrust, the axial clearance increases each time the compressor unloads. Over time, this clearance will increase enough to permit a dramatic rise in the impact energy created by axial thrust during the transient from loaded to unloaded conditions. In extreme cases, the energy can be enough to physically push the rotor assembly through the compressor housing.

Compression ratio and maximum inlet temperature determine the maximum discharge temperature of these compressors. Discharge temperatures must be limited to prevent excessive distortion between the inlet and discharge ends of the casing and rotor expansion. High-pressure units are water-jacketed to obtain uniform casing temperature. Rotors also may be cooled to permit a higher operating temperature.

If either casing distortion or rotor expansion occur, the clearance between the rotating parts will decrease and metal-to-metal contact will occur. Since the rotors typically rotate at speeds between 3,600 and 10,000 rpm, metal-to-metal contact normally results in instantaneous, catastrophic compressor failure.

Changes in differential pressures can be caused by variations in either inlet or discharge conditions (i.e., temperature, volume, or pressure). Such changes can cause the rotors to become unstable and change the load zones in the shaft-support bearings. The result is premature wear and/or failure of the bearings.

Always install a relief valve that is capable of bypassing the full-load capacity of the compressor between its discharge port and the first isolation valve. Since helical-lobe compressors are less tolerant to over-pressure operation, safety valves are usually set within 10% of absolute discharge pressure, or 5 psi, whichever is lower.

Liquid-Seal Ring The liquid-ring, or liquid-piston, compressor is shown in Figure 12.8. It has a rotor with multiple forward-turned blades that rotate about a central cone that contains inlet and discharge ports. Liquid is trapped

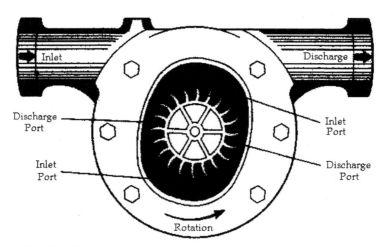

Figure 12.8 Liquid-seal ring rotary air compressor.

between adjacent blades, which drive the liquid around the inside of an elliptical casing. As the rotor turns, the liquid face moves in and out of this space because of the casing shape, creating a liquid piston. Porting in the central cone is built-in and fixed and there are no valves.

Compression occurs within the pockets or chambers between the blades before the discharge port is uncovered. Since the port location must be designed and built for a specific compression ratio, it tends to operate above or below the design pressure (refer back to Figure 12.6).

Liquid-ring compressors are cooled directly rather than by jacketed casing walls. The cooling liquid is fed into the casing, where it comes into direct contact with the gas being compressed. The excess liquid is discharged with the gas. The discharged mixture is passed through a conventional baffle or centrifugal-type separator to remove the free liquid. Because of the intimate contact of gas and liquid, the final discharge temperature can be held close to the inlet cooling water temperature. However, the discharge gas is saturated with liquid at the discharge temperature of the liquid.

The amount of liquid passed through the compressor is not critical and can be varied to obtain the desired results. The unit will not be damaged if a large quantity of liquid inadvertently enters its suction port.

Lubrication is required only in the bearings, which are generally located external to the casing. The liquid itself acts as a lubricant, sealing medium, and coolant for the stuffing boxes.

Performance

Performance of a rotary positive-displacement compressor can be evaluated by using the same criteria as used with a positive-displacement pump. Because these are constant-volume machines, performance is determined by rotation speed, internal slip, and total backpressure on the compressor.

The volumetric output of rotary positive-displacement compressors can be controlled by speed changes. The slower the compressor turns, the lower its output volume. This feature permits the use of these compressors in load-following applications. However, care must be taken to prevent sudden radical changes in speed.

Internal slip is simply the amount of gas that can flow through internal clearances from the discharge back to the inlet. Obviously, internal wear will increase internal slip.

Discharge pressure is relatively constant regardless of operating speed. With the exceptions of slight pressure variations caused by atmospheric changes and backpressure, a rotary positive-displacement compressor will provide a fixed discharge pressure. Backpressure, which is caused by restrictions in the discharge piping or demand from users of the compressed air or gas, can have a serious impact on compressor performance.

If backpressure is too low or demand too high, the compressor will be unable to provide sufficient volume or pressure to the downstream systems. In this instance, the discharge pressure will be noticeably lower than designed.

If the backpressure is too high or demand too low, the compressor will generate a discharge pressure higher than designed. It will continue to compress the air or gas until it reaches the unload setting on the system's relief valve or until the brake horsepower required exceeds the maximum horsepower rating of the driver.

Installation

Installation requirements for rotary positive-displacement compressors are similar to those for any rotating machine. Review the installation requirements for centrifugal pumps and compressors for foundation, pressure-relief, and other requirements. As with centrifugal compressors, rotary positive-displacement compressors must be fitted with pressure-relief devices to limit the discharge or interstage pressures to a safe maximum for the equipment served.

In applications in which demand varies, rotary positive-displacement compressors require a downstream receiver tank or reservoir that minimizes the load-unload cycling frequency of the compressor. The receiver tank should have sufficient volume to permit acceptable unload frequencies for the compressor. Refer to the vendor's O&M manual for specific receiver-tank recommendations.

Operating Methods

All compressor types have moving parts, high noise levels, high pressures, and high-temperature cylinder and discharge-piping surfaces. Refer to Chapter 4, which discusses compressor safety issues in general. Rotary positive-displacement compressors should be operated as base-loaded units. They are especially sensitive to the repeated start-stop operation required by load-following applications. Generally, rotary positive-displacement compressors are designed to unload about every 6 to 8 hours. This unload cycle is needed to dissipate the heat generated by the compression process. If the unload frequency is too great, these compressors have a high probability of failure.

There are several primary operating control inputs for rotary positive-displacement compressors. These control inputs are discharge pressure, pressure fluctuations, and unloading frequency.

Discharge Pressure This type of compressor will continue to compress the air volume in the down-stream system until (1) some component in the system fails, (2) the brake horsepower exceeds the driver's capacity, or (3) a safety valve opens. Therefore the operator's primary control input should be the compressor's discharge pressure. If the discharge pressure is below the design point, it is a clear indicator that the total downstream demand is greater than the unit's capacity. If the discharge pressure is too high, the demand is too low and excessive unloading will be required to prevent failure.

Pressure Fluctuations Fluctuations in the inlet and discharge pressures indicate potential system problems that may adversely affect performance and reliability. Pressure fluctuations are generally caused by changes in the ambient environment, turbulent flow, or restrictions caused by partially blocked inlet filters. Any of these problems will result in performance and reliability problems if not corrected.

Unloading Frequency The unloading function in rotary positive-displacement compressors is automatic and not under operator control. Generally, a set of limit switches, one monitoring internal temperature and one monitoring discharge pressure, is used to trigger the unload process. By design, the limit switch that monitors the compressor's internal temperature is the primary control. The secondary control, or discharge-pressure switch, is a fail-safe design to prevent overloading the compressor.

Depending on design, rotary positive-displacement compressors have an internal mechanism designed to minimize the axial thrust caused by the instantaneous change from fully loaded to unloaded operating conditions. In some designs, a balancing piston is used to absorb the rotor's thrust during this transient. In others, oversized thrust bearings are used.

Regardless of the mechanism used, none provides complete protection from the damage imparted by the transition from load to no-load conditions. However, as long as the unload frequency is within design limits, this damage will not adversely affect the compressor's useful operating life or reliability. However, an unload frequency greater than that accommodated in the design will reduce the useful life of the compressor and may lead to premature, catastrophic failure.

Operating practices should minimize, as much as possible, the unload frequency of these compressors. Installation of a receiver tank and modification of user-demand practices are the most effective solutions to this type of problem.

RECIPROCATING

Reciprocating compressors are widely used by industry and are offered in a wide range of sizes and types. They vary from units requiring less than 1 Hp to more than 12,000 Hp. Pressure capabilities range from low vacuums at intake to special compressors capable of 60,000 psig or higher.

Reciprocating compressors are classified as constant-volume, variable-pressure machines. They are the most efficient type of compressor and can be used for partial-load, or reduced-capacity, applications.

Because of the reciprocating pistons and unbalanced rotating parts, the unit tends to shake. Therefore it is necessary to provide a mounting that stabilizes the installation. The extent of this requirement depends on the type and size of the compressor.

Because reciprocating compressors should be supplied with clean gas, inlet filters are recommended in all applications. They cannot satisfactorily handle liquids entrained in the gas, although vapors are no problem if condensation within the cylinders does not take place. Liquids will destroy the lubrication and cause excessive wear.

Reciprocating compressors deliver a pulsating flow of gas that can damage downstream equipment or machinery. This is sometimes a disadvantage, but pulsation dampers can be used to alleviate the problem.

Configuration

Certain design fundamentals should be clearly understood before analyzing the operating condition of reciprocating compressors. These fundamentals include frame and running gear, inlet and discharge valves, cylinder cooling, and cylinder orientation.

Frame and Running Gear

Two basic factors guide frame and running gear design. The first factor is the maximum horsepower to be transmitted through the shaft and running gear to the cylinder pistons. The second factor is the load imposed on the frame parts by the pressure differential between the two sides of each piston. This is often called *pin load* because this full force is directly exerted on the crosshead and crankpin. These two factors determine the size of bearings, connecting rods, frame, and bolts that must be used throughout the compressor and its support structure.

Cylinder Design

Compression efficiency depends entirely on the design of the cylinder and its valves. Unless the valve area is sufficient to allow gas to enter and leave the cylinder without undue restriction, efficiency cannot be high. Valve placement for free flow of the gas in and out of the cylinder is also important.

Both efficiency and maintenance are influenced by the degree of cooling during compression. The method of cylinder cooling must be consistent with the service intended.

The cylinders and all the parts must be designed to withstand the maximum application pressure. The most economical materials that will give the proper strength and the longest service under the design conditions are generally used.

Inlet and Discharge Valves

Compressor valves are placed in each cylinder to permit one-way flow of gas, either into or out of the cylinder. There must be one or more valve(s) for inlet and discharge in each compression chamber.

Each valve opens and closes once for each revolution of the crankshaft. The valves in a compressor operating at 700 rpm for 8 hours per day and 250 days per year will have cycled (i.e., opened and closed) 42,000 times per hour, 336,000 times per day, or 84 million times in a year. The valves have less than $\frac{1}{10}$ of a second to open, let the gas pass through, and close. They must cycle with a minimum of resistance for minimum power consumption. However, the valves must have minimal clearance to prevent excessive expansion and reduced volumetric efficiency. They must be tight under extreme pressure and temperature conditions. Finally, the valves must be durable under many kinds of abuse.

There are four basic valve designs used in these compressors: finger, channel, leaf, and annular ring. Within each class there may be variations in design, depending on operating speed and size of valve required.

Finger Figure 12.9 is an exploded view of a typical finger valve. These valves are used for smaller, air-cooled compressors. One end of the finger is fixed and the opposite end lifts when the valve opens.

Channel The channel valve shown in Figure 12.10 is widely used in mid- to large-sized compressors. This valve uses a series of separate stainless steel channels. As explained in the figure, this is a cushioned valve, which adds greatly to its life.

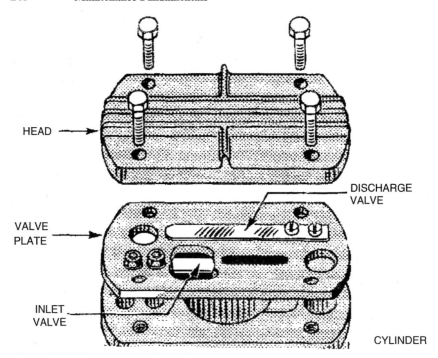

HEAD

VALVE
PLATE

DISCHARGE
VALVE

INLET
VALVE

CYLINDER

Figure 12.9 Finger valve configuration.

Leaf The leaf valve (Figure 12.11) has a configuration somewhat like the channel valve. It is made of flat-strip steel that opens against an arched stop plate. This results in valve flexing only at its center with maximum lift. The valve operates as its own spring.

Annular Ring Figure 12.12 shows exploded views of typical inlet and discharge annular-ring valves. The valves shown have a single ring, but larger sizes may have two or three rings. In some designs, the concentric rings are tied into a single piece by bridges.

The springs and the valve move into a recess in the stop plate as the valve opens. Gas that is trapped in the recess acts as a cushion and prevents slamming. This eliminates a major source of valve and spring breakage. The valve shown was the first cushioned valve built.

Cylinder Cooling

Cylinder heat is produced by the work of compression plus friction, which is caused by the action of the piston and piston rings on the cylinder wall and

Valve Closed: A tight seat is formed
without slamming or friction, so seat wear
is at a minimum. Both channel and spring
are precision made to assure a perfect fit.
A gas space is formed between the bowed
spring and the flat channel.

Valve Opening: Channel lifts straight up
in the guides without flexing. Opening is
even over the full length of the port, giving
uniform air velocity without turbulence.
Cushioning is effected by the compression
and escape of the gas between spring and
channel.

Valve Wide Open: Gas trapped between
spring and channel has been compressed
and in escaping has allowed channel to
float in its stop.

Figure 12.10 Channel valve configuration.

packing on the rod. The amount of heat generated can be considerable, particularly when moderate to high compression ratios are involved. This can result in undesirably high operating temperatures.

Most compressors use some method to dissipate a portion of this heat to reduce the cylinder wall and discharge gas temperatures. The following are advantages of cylinder cooling:

- Lowering cylinder wall and cylinder head temperatures reduces loss of capacity and horsepower per unit volume caused by suction gas preheating during inlet stroke. This results in more gas in the cylinder for compression.
- Reducing cylinder wall and cylinder head temperatures removes more heat from the gas during compression, lowering its final temperature and reducing the power required.

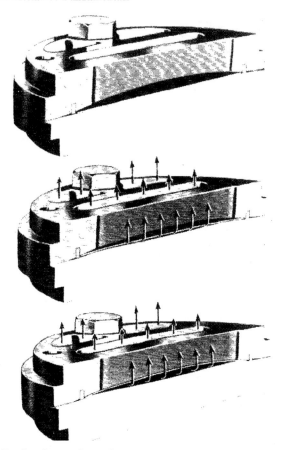

Figure 12.11 Leaf spring configuration.

- Reducing the gas temperature and that of the metal surrounding the valves results in longer valve service life and reduces the possibility of deposit formation.
- Reduced cylinder wall temperature promotes better lubrication, resulting in longer life and reduced maintenance.
- Cooling, particularly water cooling, maintains a more even temperature around the cylinder bore and reduces warpage.

Cylinder Orientation

Orientation of the cylinders in a multi-stage or multi-cylinder compressor directly affects the operating dynamics and vibration level. Figure 12.13 illustrates a typical three-piston, air-cooled compressor. Since three pistons are oriented

INLET   DISCHARGE

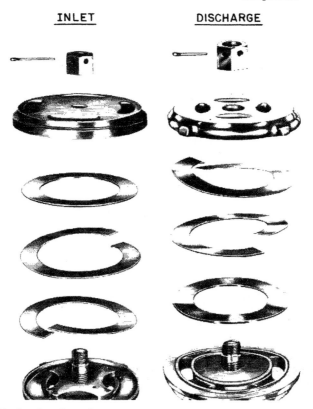

Figure 12.12 Annular-ring valves.

within a 120-degree arc, this type of compressor generates higher vibration levels than the opposed piston compressor illustrated in Figure 12.14.

Performance

Reciprocating-compressor performance is governed almost exclusively by operating speed. Each cylinder of the compressor will discharge the same volume, excluding slight variations caused by atmospheric changes, at the same discharge pressure each time it completes the discharge stroke. As the rotation speed of the compressor changes, so does the discharge volume. The only other variables that affect performance are the inlet-discharge valves, which control flow into and out of each cylinder. Although reciprocating compressors can use a variety of valve designs, it is crucial that the valves perform reliably. If they are damaged and fail to operate at the proper time or do not seal properly, overall compressor performance will be substantially reduced.

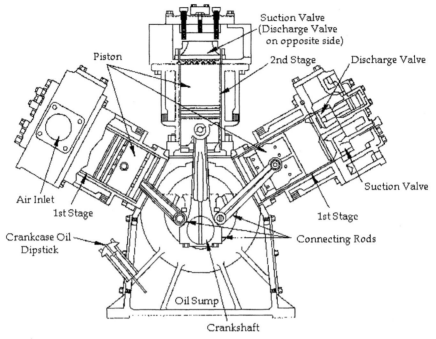

Figure 12.13 Three-piston compressor generates higher vibration levels.

Installation

A carefully planned and executed installation is extremely important and makes compressor operation and maintenance easier and safer. Key components of a compressor installation are location, foundation, and piping.

Location The preferred location for any compressor is near the center of its load. However, the choice is often influenced by the cost of supervision, which can vary by location. The ongoing cost of supervision may be less expensive at a less-optimum location, which can offset the cost of longer piping.

A compressor will always give better, more reliable service when enclosed in a building that protects it from cold, dusty, damp, and corrosive conditions. In certain locations it may be economical to use a roof only, but this is not recommended unless the weather is extremely mild. Even then, it is crucial to prevent rain and wind-blown debris from entering the moving parts. Subjecting a compressor to adverse inlet conditions will dramatically reduce reliability and significantly increase maintenance requirements.

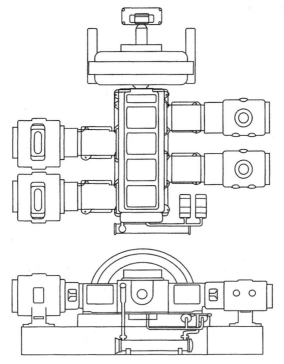

Figure 12.14 Opposed-piston compressor balances piston forces.

Ventilation around a compressor is vital. On a motor-driven, air-cooled unit, the heat radiated to the surrounding air is at least 65% of the power input. On a water-jacketed unit with an aftercooler and outside receiver, the heat radiated to the surrounding air may be 15–25% of the total energy input, which is still a substantial amount of heat. Positive outside ventilation is recommended for any compressor room where the ambient temperature may exceed 104°F.

Foundation Because of the alternating movement of pistons and other components, reciprocating compressors often develop a shaking that alternates in direction. This force must be damped and contained by the mounting. The foundation also must support the weight load of the compressor and its driver.

There are many compressor arrangements and the net magnitude of the moments and forces developed can vary a great deal among them. In some cases, they are partially or completely balanced within the compressors themselves. In others, the foundation must handle much of the force. When complete balance is possible, reciprocating compressors can be mounted on a foundation just large and rigid enough to carry the weight and maintain alignment.

However, most reciprocating compressors require larger, more massive foundations than other machinery.

Depending on size and type of unit, the mounting may vary from simply bolting to the floor, to attaching to a massive foundation designed specifically for the application. A proper foundation must (1) maintain the alignment and level of the compressor and its driver at the proper elevation and (2) minimize vibration and prevent its transmission to adjacent building structures and machinery. There are five steps to accomplish the first objective:

1. The safe weight-bearing capacity of the soil must not be exceeded at any point on the foundation base.
2. The load to the soil must be distributed over the entire area.
3. The size and proportion of the foundation block must be such that the resultant vertical load caused by the compressor, block, and any unbalanced force falls within the base area.
4. The foundation must have sufficient mass and weight-bearing area to prevent its sliding on the soil because of unbalanced forces.
5. Foundation temperature must be uniform to prevent warping.

Bulk is not usually the complete solution to foundation problems. A certain weight is sometimes necessary, but soil area is usually of more value than foundation mass.

Determining whether two or more compressors should have separate or single foundations depends on the compressor type. A combined foundation is recommended for reciprocating units since the forces from one unit usually will partially balance out the forces from the others. In addition, the greater mass and surface area in contact with the ground damps foundation movement and provides greater stability.

Soil quality may vary seasonally, and such conditions must be carefully considered in the foundation design. No foundation should rest partially on bedrock and partially on soil; it should rest entirely on one or the other. If placed on the ground, make sure that part of the foundation does not rest on soil that has been disturbed. In addition, pilings may be necessary to ensure stability.

Piping Piping should easily fit the compressor connections without needing to spring or twist it to fit. It must be supported independently of the compressor and anchored, as necessary, to limit vibration and to prevent expansion strains. Improperly installed piping may distort or pull the compressor's cylinders or casing out of alignment.

Air Inlet

The intake pipe on an air compressor should be as short and direct as possible. If the total run of the inlet piping is unavoidably long, the diameter should be increased. The pipe size should be greater than the compressor's air-inlet connection.

Cool inlet air is desirable. For every 5°F of ambient air temperature reduction, the volume of compressed air generated increases by 1% with the same power consumption. This increase in performance is due to the greater density of the intake air.

It is preferable for the intake air to be taken from outdoors. This reduces heating and air conditioning costs and, if properly designed, has fewer contaminants. However, the intake piping should be a minimum of 6 ft. above the ground and be screened or, preferably, filtered. An air inlet must be free of steam and engine exhausts. The inlet should be hooded or turned down to prevent the entry of rain or snow. It should be above the building eaves and several feet from the building.

Discharge

Discharge piping should be the full size of the compressor's discharge connection. The pipe size should not be reduced until the point along the pipeline is reached where the flow has become steady and non-pulsating. With a reciprocating compressor, this is generally beyond the aftercooler or the receiver. Pipes to handle non-pulsating flow are sized by normal methods and long-radius bends are recommended. All discharge piping must be designed to allow adequate expansion loops or bends to prevent undue stresses at the compressor.

Drainage

Before piping is installed, the layout should be analyzed to eliminate low points where liquid could collect and to provide drains where low points cannot be eliminated. A regular part of the operating procedure must be the periodic drainage of low points in the piping and separators, as well as inspection of automatic drain traps.

Pressure-Relief Valves

All reciprocating compressors must be fitted with pressure relief devices to limit the discharge or interstage pressures to a safe maximum for the equipment served. Always install a relief valve that is capable of bypassing the full-load capacity of the compressor between its discharge port and the first isolation valve. The safety valves should be set to open at a pressure slightly higher than the normal discharge-pressure rating of the compressor. For standard 100–115 psig two-stage air compressors, safety valves are normally set at 125 psig.

The pressure-relief safety valve is normally situated on top of the air reservoir, and there must be no restriction on its operation. The valve is usually of the "huddling chamber" design in which the static pressure acting on its disk area causes it to open. Figure 12.15 illustrates how such a valve functions. As the valve pops, the air space within the huddling chamber between the seat and blowdown ring fills with pressurized air and builds up more pressure on the roof of the disk holder. This temporary pressure increases the upward thrust against the spring, causing the disk and its holder to fully pop open.

Once a predetermined pressure drop (i.e., blowdown) occurs, the valve closes with a positive action by trapping pressurized air on top of the disk holder. Raising or

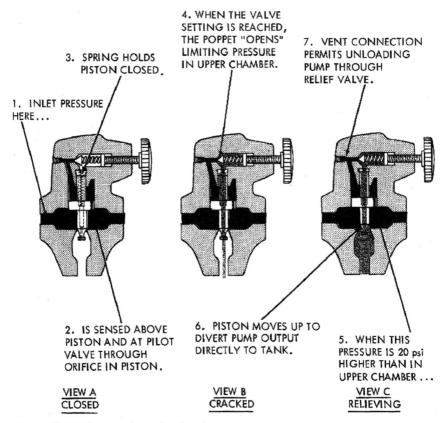

Figure 12.15 How a safety valve functions.

lowering the blowdown ring adjusts the pressure-drop setpoint. Raising the ring increases the pressure-drop setting, while lowering it decreases the setting.

Operating Methods

Compressors can be hazardous to work around because they have moving parts. Ensure that clothing is kept away from belt drives, couplings, and exposed shafts. In addition, high-temperature surfaces around cylinders and discharge piping are exposed. Compressors are notoriously noisy, so ear protection should be worn. These machines are used to generate high-pressure gas so, when working around them, it is important to wear safety glasses and to avoid searching for leaks with bare hands. High-pressure leaks can cause severe friction burns.

TROUBLESHOOTING

Compressors can be divided into three classifications: centrifugal, rotary, and reciprocating. This section identifies the common failure modes for each.

CENTRIFUGAL

The operating dynamics of centrifugal compressors are the same as for other centrifugal machine-trains. The dominant forces and vibration profiles are typically identical to pumps or fans. However, the effects of variable load and other process variables (e.g., temperatures, inlet/discharge pressure, etc.) are more pronounced than in other rotating machines. Table 12.1 identifies the common failure modes for centrifugal compressors.

Aerodynamic instability is the most common failure mode for centrifugal compressors. Variable demand and restrictions of the inlet-air flow are common sources of this instability. Even slight variations can cause dramatic changes in the operating stability of the compressor.

Entrained liquids and solids also can affect operating life. When dirty air must be handled, open-type impellers should be used. An open design provides the ability to handle a moderate amount of dirt or other solids in the inlet-air supply. However, inlet filters are recommended for all applications, and controlled liquid injection for cleaning and cooling should be considered during the design process.

ROTARY-TYPE, POSITIVE DISPLACEMENT

Table 12.2 lists the common failure modes of rotary-type, positive-displacement compressors. This type of compressor can be grouped into two types, sliding vane and rotary screw.

Table 12.1 Common Failure Modes of Centrifugal Compressors

THE CAUSES	Excessive Vibration	Compressor Surges	Loss of Discharge Pressure	Low Lube Oil Pressure	Excessive Bearing Oil Drain Temp.	Units Do Not Stay in Alignment	Persistent Unloading	Water in Lube Oil	Motor Trips
							THE PROBLEM		
Bearing Lube Oil Orifice Missing or Plugged					●				
Bent Rotor (Caused by Uneven Heating and Cooling)	●						●		
Build up of Deposits on Diffuser		●							
Build up of Deposits on Rotor	●	●							
Change in System Resistance		●							●
Clogged Oil Strainer/Filter					●				
Compressor Not Up To Speed			●						
Condensate in Oil Reservoir								●	
Damaged Rotor	●								
Dry Gear Coupling	●								
Excessive Bearing Clearance	●								
Excessive Inlet Temperature			●						
Failure of Both Main and Auxiliary Oil Pumps				●					
Faulty Temperature Gauge or Switch					●	●			●
Improperly Assembled Parts	●						●		●
Incorrect Pressure Control Valve Setting				●					
Insufficient Flow		●							
Leak in Discharge Piping			●						
Leak in Lube Oil Cooler Tubes or Tube Sheet								●	
Leak in Oil Pump Suction Piping				●					
Liquid "Slugging"	●						●		

Table 12.1 (continued)

THE CAUSES	Excessive Vibration	Compressor Surges	Loss of Discharge Pressure	Low Lube Oil Pressure	Excessive Bearing Oil Drain Temp.	Units Do Not Stay in Alignment	Persistent Unloading	Water in Lube Oil	Motor Trips
Loose or Broken Bolting	•								
Loose Rotor Parts	•								
Oil Leakage				•					
Oil Pump Suction Plugged				•					
Oil Reservoir Low Level				•					
Operating at Low Speed w/o Auxiliary Oil Pump				•					
Operating in Critical Speed Range	•								
Operating in Surge Region	•								
Piping Strain	•					•	•	•	•
Poor Oil Condition					•				
Relief Valve Improperly Set or Stuck Open				•					
Rotor Imbalance	•						•		
Rough Rotor Shaft Journal Surface					•		•		•
Shaft Misalignment	•					•			
Sympathetic Vibration	•						•	•	
Vibration					•				
Warped Foundation or Baseplate							•		•
Wiped or Damaged Bearings					•				•
Worn or Damaged Coupling	•								

Table 12.2 Common Failure Modes of Rotary-Type, Positive-Displacement Compressors

THE CAUSES	No Air/Gas Delivery	Insufficient Discharge Pressure	Insufficient Capacity	Excessive Wear	Excessive Heat	Excessive Vibration and Noise	Excessive Power Demand	Motor Trips	Elevated Motor Temperature	Elevated Air/Gas Temperature
Air Leakage into Suction Piping or Shaft Seal		•	•			•				
Coupling Misaligned				•	•	•	•		•	
Excessive Discharge Pressure			•	•		•	•	•		•
Excessive Inlet Temperature/Moisture			•							
Insufficient Suction Air/Gas Supply		•	•	•		•			•	
Internal Component Wear	•	•	•							
Motor or Driver Failure	•									
Pipe Strain on Compressor Casing				•	•	•	•		•	
Relief Valve Stuck Open or Set Wrong		•	•							
Rotating Element Binding				•	•	•	•	•	•	
Solids or Dirt in Inlet Air/Gas Supply				•						
Speed Too Low		•	•						•	
Suction Filter or Strainer Clogged	•	•	•				•		•	
Wrong Direction of Rotation	•	•							•	

SLIDING-VANE COMPRESSORS

Sliding-vane·compressors have the same failure modes as vane-type pumps. The dominant components in their vibration profile are running speed, vane-pass frequency, and bearing-rotation frequencies. In normal operation, the dominant energy is at the shaft's running speed. The other frequency components are at much lower energy levels. Common failures of this type of compressor occur with shaft seals, vanes, and bearings.

Shaft Seals

Leakage through the shaft's seals should be checked visually once a week or as part of every data-acquisition route. Leakage may not be apparent from the outside of the gland. If the fluid is removed through a vent, the discharge should be configured for easy inspection. Generally, more leakage than normal is the signal to replace a seal. Under good conditions, they have a normal life of 10,000 to 15,000 hours and should routinely be replaced when this service life has been reached.

Vanes

Vanes wear continuously on their outer edges and, to some degree, on the faces that slide in and out of the slots. The vane material is affected somewhat by prolonged heat, which causes gradual deterioration. Typical life expectancy of vanes in 100-psig service is about 16,000 hours of operation. For low-pressure applications, life may reach 32,000 hours.

Replacing vanes before they break is extremely important. Breakage during operation can severely damage the compressor, which requires a complete overhaul and realignment of heads and clearances.

Bearings

In normal service, bearings have a relatively long life. Replacement after about 6 years of operation is generally recommended. Bearing defects are usually displayed in the same manner in a vibration profile as for any rotating machine-train. Inner and outer race defects are the dominant failure modes, but roller spin also may contribute to the failure.

Rotary Screw

The most common reason for compressor failure or component damage is process instability. Rotary-screw compressors are designed to deliver a constant volume and pressure of air or gas. These units are extremely susceptible to any change in either inlet or discharge conditions. A slight variation in pressure, temperature, or volume can result in instantaneous failure. The following are used as indices of instability and potential problems: rotor mesh, axial movement, thrust bearings, and gear mesh.

Rotor Mesh

In normal operation, the vibration energy generated by male and female rotor meshing is very low. As the process becomes unstable, the energy caused by the rotor meshing frequency increases, with both the amplitude of the meshing frequency and the width of the peak increasing. In addition, the noise floor surrounding the meshing frequency becomes more pronounced. This white noise is similar to that observed in a cavitating pump or unstable fan.

Axial Movement

The normal tendency of the rotors and helical timing gears is to generate axial shaft movement, or thrusting. However, the extremely tight clearances between the male and female rotors do not tolerate any excessive axial movement and, therefore, axial movement should be a primary monitoring parameter. Axial measurements are needed from both rotor assemblies. If there is any increase in the vibration amplitude of these measurements, it is highly probable that the compressor will fail.

Thrust Bearings

While process instability can affect both the fixed and float bearings, the thrust bearing is more likely to show early degradation as a result of process instability or abnormal compressor dynamics. Therefore these bearings should be monitored closely, and any degradation or hint of excessive axial clearance should be corrected immediately.

Gear Mesh

The gear mesh vibration profile also provides an indication of prolonged compressor instability. Deflection of the rotor shafts changes the wear pattern on the helical gear sets. This change in pattern increases the backlash in the gear mesh, results in higher vibration levels, and increases thrusting.

RECIPROCATING, POSITIVE DISPLACEMENT

Reciprocating compressors have a history of chronic failures that include valves, lubrication system, pulsation, and imbalance. Table 12.3 identifies common failure modes and causes for this type of compressor.

Like all reciprocating machines, reciprocating compressors normally generate higher levels of vibration than centrifugal machines. In part, the increased level of vibration is caused by the impact as each piston reaches top dead center and bottom dead center of its stroke. The energy levels also are influenced by the unbalanced forces generated by non-opposed pistons and looseness in the piston rods, wrist pins, and journals of the compressor. In most cases, the dominant vibration frequency is the second harmonic (2X) of the main crankshaft's rotating speed. Again, this results from the impact that occurs when each piston changes directions (i.e., two impacts occur during one complete crankshaft rotation).

Valves

Valve failure is the dominant failure mode for reciprocating compressors. Because of their high cyclic rate, which exceeds 80 million cycles per year, inlet and discharge valves tend to work harden and crack.

Lubrication System

Poor maintenance of lubrication-system components, such as filters and strainers, typically causes premature failure. Such maintenance is crucial to reciprocating compressors because they rely on the lubrication system to provide a uniform oil film between closely fitting parts (e.g., piston rings and the cylinder wall). Partial or complete failure of the lube system results in catastrophic failure of the compressor.

Pulsation

Reciprocating compressors generate pulses of compressed air or gas that are discharged into the piping that transports the air or gas to its point(s) of use. This pulsation often generates resonance in the piping system and pulse impact (i.e., standing waves) can severely damage other machinery connected to the compressed-air system. While this behavior does not cause the compressor to fail, it must be prevented to protect other plant equipment. Note, however, that most compressed-air systems do not use pulsation dampers.

Each time the compressor discharges compressed air, the air tends to act like a compression spring. Because it rapidly expands to fill the discharge piping's available volume, the pulse of high-pressure air can cause serious damage. The pulsation wavelength, λ, from a compressor having a double-acting piston design can be determined by:

$$\lambda = \frac{60a}{2n} = \frac{34,050}{n}$$

CRANK ARRANGEMENTS	FORCES		COUPLES	
	PRIMARY	SECONDARY	PRIMARY	SECONDARY
SINGLE CRANK 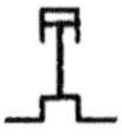	F'WITHOUT COUNTERWTS. 0.5 F' WITH COUNTERWTS.	F"	NONE	NONE
TWO CRANKS AT 180° IN LINE CYLINDERS 	ZERO	2F"	F'D WITHOUT COUNTERWTS. $\frac{F'D}{2}$ WIDTH COUNTERWTS.	NONE
OPPOSED CYLINDERS 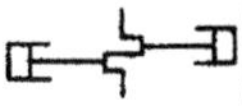	ZERO	ZERO	NIL	NIL
TWO CRANKS AT 90° 	141 F' WITHOUT COUNTERWTS. 0.707 F' WITH COUNTERWTS.	ZERO	707F'D WITHOUT COUNTERWTS. 0.354F'D WITH COUNTERWTS.	F"D
TWO CYLINDERS ON ONE CRANK CYLINDERS AT 90° 	F'WITHOUT COUNTERWTS. ZERO WITHOUT COUNTERWTS.	141F"	NIL	NIL
TWO CYLINDERS ON ONE CRANK OPPOSED CYLINDERS 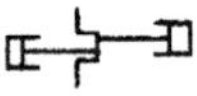	2F' WITHOUT COUNTERWTS. F' WITH COUNTERWTS.	ZERO	NONE	NIL
THREE CRANKS AT 120° 	ZERO	ZERO		3.40F"D
FOUR CYLINDERS CRANKS AT 180° 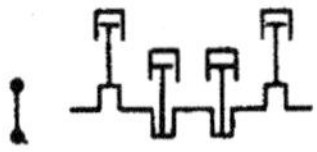	ZERO	4F"	ZERO	ZERO
CRANKS AT 90° 	ZERO	ZERO	1.41F'D WITHOUT COUNTERWTS. 0.707F'D WITH COUNTERWTS.	4.0F"D
SIX CYLINDERS 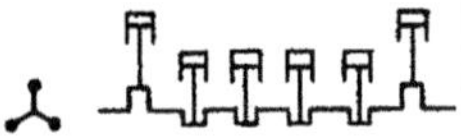	ZERO	ZERO	ZERO	ZERO

F' = PRIMARY INERTIA FORCE IN LBS.
F' = .000028RN^{2}W
F" = SECONDARY INERTIA FORCE IN LBS.

$$F'' = \frac{R}{L}F'$$

R = CRANK RADIUS, INCHES
N = R.P.M.
W = RECIPROCATING WEIGHT OF ONE CYLINDER, LBS
L = LENGTH OF CONNECTING ROD, INCHES
D = CYLINDER CENTER DISTANCE

Figure 12.16 Unbalanced inertial forces and couples for various reciprocating compressors.

where

λ = Wavelength, feet
a = Speed of sound = 1,135 feet/second
n = Compressor speed, revolutions/minute

For a double-acting piston design, a compressor running at 1,200 rpm will generate a standing wave of 28.4 ft. In other words, a shock load equivalent to the discharge pressure will be transmitted to any piping or machine connected to the discharge piping and located within 28 ft. of the compressor. Note that for a single-acting cylinder, the wavelength will be twice as long.

Imbalance

Compressor inertial forces may have two effects on the operating dynamics of a reciprocating compressor, affecting its balance characteristics. The first is a force in the direction of the piston movement, which is displayed as impacts in a vibration profile as the piston reaches top and bottom dead center of its stroke. The second effect is a couple, or moment, caused by an offset between the axes of two or more pistons on a common crankshaft. The inter-relationship and magnitude of these two effects depend on such factors as: (1) number of cranks, (2) longitudinal and angular arrangement, (3) cylinder arrangement, and (4) amount of counter balancing possible. Two significant vibration periods result, the primary at the compressor's rotation speed (X) and the secondary at 2X.

Although the forces developed are sinusoidal, only the maximum (i.e., the amplitude) is considered in the analysis. Figure 12.16 shows relative values of the inertial forces for various compressor arrangements.

13

CONTROL VALVES

Control valves can be grouped into two major classifications: process and fluid power. Process valves control the flow of gases and liquids through a process system. Fluid-power valves control pneumatic or hydraulic systems.

PROCESS

Process-control valves are available in a variety of sizes, configurations, and materials of construction. Generally, this type of valve is classified by its internal configuration.

Configuration

The device used to control flow through a valve varies with its intended function. The more common types are ball, gate, butterfly, and globe valves.

Ball

Ball valves (Figure 13.1) are simple shut-off devices that use a ball to stop and start the flow of fluid downstream of the valve. As the valve stem turns to the open position, the ball rotates to a point where part or the entire hole machined through the ball is in line with the valve-body inlet and outlet. This allows fluid to pass through the valve. When the ball rotates so that the hole is perpendicular to the flow path, the flow stops.

Most ball valves are quick-acting and require a 90-degree turn of the actuator lever to fully open or close the valve. This feature, coupled with the turbulent

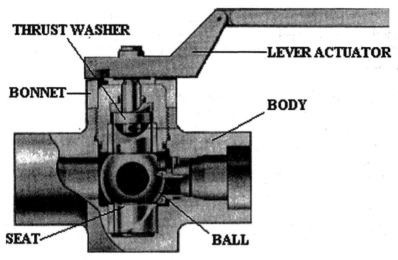

THRUST WASHER

LEVER ACTUATOR

BONNET

BODY

SEAT

BALL

Figure 13.1 Ball valve.

flow generated when the ball opening is only partially open, limits the use of the ball valve. Use should be limited to strictly an "on-off" control function (i.e., fully open or fully closed) because of the turbulent-flow condition and severe friction loss when in the partially open position. They should not be used for throttling or flow-control applications.

Ball valves used in process applications may incorporate a variety of actuators to provide direct or remote control of the valve. The more common actuators are either manual or motor-operated. Manual values have a hand wheel or lever attached directly or through a gearbox to the valve stem. The valve is opened or closed by moving the valve stem through a 90-degree arc.

Motor-controlled valves replace the hand wheel with a fractional horsepower motor that can be controlled remotely. The motor-operated valve operates in exactly the same way as the manually operated valve.

Gate

Gate valves are used when straight line, laminar fluid flow, and minimum restrictions are needed. These valves use a wedge-shaped sliding plate in the valve body to stop, throttle, or permit full flow of fluids through the valve. When the valve is wide open, the gate is completely inside the valve bonnet. This leaves the flow passage through the valve fully open with no flow restrictions, allowing little or no pressure drop through the valve.

Gate valves are not suitable for throttling the flow volume unless specifically authorized for this application by the manufacturer. They generally are not suitable because the flow of fluid through a partially open gate can cause extensive damage to the valve.

Gate valves are classified as either rising stem or non rising stem. In the non-rising-stem valve, which is shown in Figure 13.2, the stem is threaded into the gate. As the hand wheel on the stem is rotated, the gate travels up or down the stem on the threads, while the stem remains vertically stationary. This type of valve will almost always have a pointer indicator threaded onto the upper end of the stem to indicate the position of the gate.

Valves with rising stems (Figure 13.3) are used when it is important to know by immediate inspection if the valve is open or closed, or when the threads exposed to the fluid could become damaged by fluid contamination. In this valve, the stem rises out of the valve bonnet when the valve is opened.

Butterfly

The butterfly valve has a disk-shaped element that rotates about a central shaft or stem. When the valve is closed, the disk face is across the pipe and blocks the flow. Depending on the type of butterfly valve, the seat may consist of a bonded resilient liner, a mechanically fastened resilient liner, an insert-type reinforced resilient liner, or an integral metal seat with an O-ring inserted around the edge of the disk.

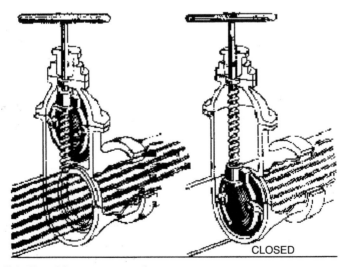

CLOSED

Figure 13.2 Non-rising-stem gate valve.

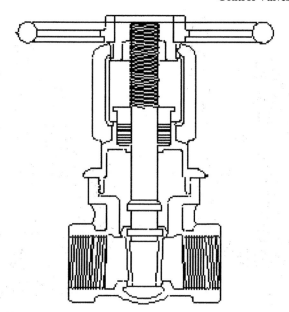

Figure 13.3 Rising stem gate valve.

As shown in Figure 13.4, both the full open and the throttled positions permit almost unrestricted flow. Therefore, this valve does not induce turbulent flow in the partially closed position. While the design does not permit exact flow-control capabilities, a butterfly valve can be used for throttling flow through the valve. In addition, these valves have the lowest pressure drop of all the conventional types. For these reasons, they are commonly used in process-control applications.

Globe

The globe valve gets its name from the shape of the valve body, although other types of valves also may have globular-shaped bodies. Figure 13.5 shows three configurations of this type of valve: straight-flow, angle-flow, and cross-flow.

A disk attached to the valve stem controls flow in a globe valve. Turning the valve stem until the disk is seated, which is illustrated in View A of Figure 13.6, closes the valve. The edge of the disk and the seat are very accurately machined to form a tight seal. It is important for globe valves to be installed with the pressure against the disk face to protect the stem packing from system pressure when the valve is shut.

While this type of valve is commonly used in the fully open or fully closed position, it also may be used for throttling. However, since the seating surface

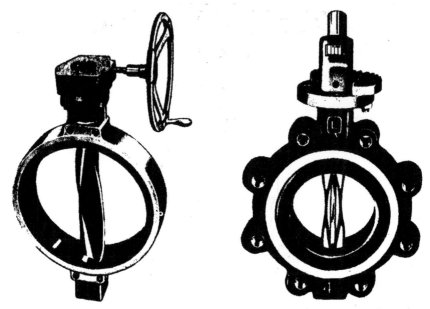

Figure 13.4 Butterfly valves provide almost unrestricted flow.

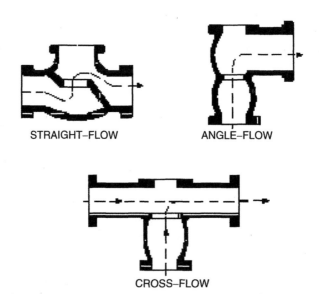

STRAIGHT–FLOW   ANGLE–FLOW

CROSS–FLOW

Figure 13.5 Three globe valve configurations: straight-flow, angle-flow, and cross-flow.

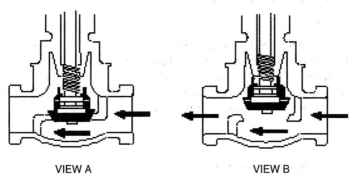

VIEW A   VIEW B

Figure 13.6 Globe valve.

is a relatively large area, it is not suitable for throttling applications in which fine adjustments are required.

When the valve is open, as illustrated in View B of Figure 13.6, the fluid flows through the space between the edge of the disk and the seat. Since the fluid flow is equal on all sides of the center of support when the valve is open, there is no unbalanced pressure on the disk to cause uneven wear. The rate at which fluid flows through the valve is regulated by the position of the disk in relation to the valve seat.

The globe valve should never be jammed in the open position. After a valve is fully opened, the hand wheel or actuating handle should be closed approximately one-half turn. If this is not done, the valve may seize in the open position, making it difficult, if not impossible, to close the valve. Many valves are damaged in this manner. Another reason to partially close a globe valve is that it can be difficult to tell if the valve is open or closed. If jammed in the open position, the stem can be damaged or broken by someone who thinks the valve is closed.

Performance

Process-control valves have few measurable criteria that can be used to determine their performance. Obviously, the valve must provide a positive seal when closed. In addition, it must provide a relatively laminar flow with minimum pressure drop in the fully open position. When evaluating valves, the following criteria should be considered: capacity rating, flow characteristics, pressure drop, and response characteristics.

Capacity Rating

The primary selection criterion of a control valve is its capacity rating. Each type of valve is available in a variety of sizes to handle most typical process-flow rates. However, proper size selection is critical to the performance characteristics of the valve and the system in which it is installed. A valve's capacity must accommodate variations in viscosity, temperature, flow rates, and upstream pressure.

Flow Characteristics

The internal design of process-control valves has a direct impact on the flow characteristics of the gas or liquid flowing through the valve. A fully open butterfly or gate valve provides a relatively straight, obstruction-free flow path. As a result, the product should not be affected.

Pressure Drop

Control-valve configuration impacts the resistance to flow through the valve. The amount of resistance, or pressure drop, will vary greatly, depending on type, size, and position of the valve's flow-control device (i.e., ball, gate, disk). Pressure-drop formulas can be obtained for all common valve types from several sources (e.g., Crane, Technical Paper No. 410).

Response Characteristics

With the exception of simple, manually controlled shut off valves, process-control valves are generally used to control the volume and pressure of gases or liquids within a process system. In most applications, valves are controlled from a remote location through the use of pneumatic, hydraulic, or electronic actuators. Actuators are used to position the gate, ball, or disk that starts, stops, directs, or proportions the flow of gas or liquid through the valve. Therefore the response characteristics of a valve are determined, in part, by the actuator. Three factors critical to proper valve operation are response time, length of travel, and repeatability.

Response Time Response time is the total time required for a valve to open or close to a specific set-point position. These positions are fully open, fully closed, and any position in between. The selection and maintenance of the actuator used to control process-control valves have a major impact on response time.

Length of Travel The valve's flow-control device (i.e., gate, ball, or disk) must travel some distance when going from one setpoint to another. With a manually operated valve, this is a relatively simple operation. The operator moves the stem lever or hand wheel until the desired position is reached. The only reasons a manually controlled valve will not position properly are mechanical wear or looseness between the lever or hand wheel and the disk, ball, or gate. For remotely controlled valves, however, there are other variables that directly affect valve travel. These variables depend on the type of actuator that is used. There are three major types of actuators: pneumatic, hydraulic, and electronic.

Pneumatic Actuators Pneumatic actuators—including diaphragms, air motors, and cylinders—are suitable for simple on-off valve applications. As long as there is enough air volume and pressure to activate the actuator, the valve can be repositioned over its full length of travel. However, when the air supply required to power the actuator is inadequate or the process-system pressure is too great, the actuator's ability to operate the valve properly is severely reduced.

A pneumatic (i.e., compressed air-driven) actuator is shown in Figure 13.7. This type is not suited for precision flow-control applications, because the compressibility of air prevents it from providing smooth, accurate valve positioning.

Hydraulic Actuators Hydraulic (i.e., fluid-driven) actuators, also illustrated in Figure 13.7, can provide a positive means of controlling process valves in most applications. Properly installed and maintained, this type of actuator can provide accurate, repeatable positioning of the control valve over its full range of travel.

Electronic Actuators Some control valves use high-torque electric motors as their actuator (Figure 13.8). If the motors are properly sized and their control circuits are maintained, this type of actuator can provide reliable, positive control over the full range of travel.

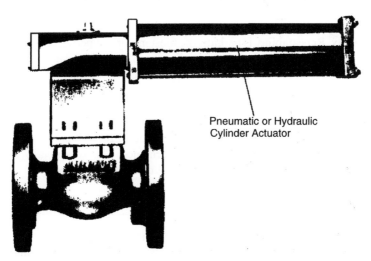

Pneumatic or Hydraulic
Cylinder Actuator

Figure 13.7 Pneumatic or hydraulic cylinders are used as actuators.

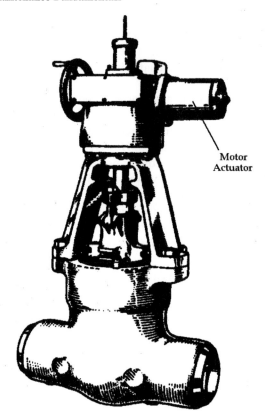

Figure 13.8 High-torque electric motors can be used as actuators.

Repeatability Repeatability is perhaps the most important performance criterion of a process-control valve. This is especially true in applications in which precise flow or pressure control is needed for optimum performance of the process system.

New process-control valves generally provide the repeatability required. However, proper maintenance and periodic calibration of the valves and their actuators are required to ensure long-term performance. This is especially true for valves that use mechanical linkages as part of the actuator assembly.

Installation

Process-control valves cannot tolerate solids, especially abrasives, in the gas or liquid stream. In applications in which high concentrations of particulates are present, valves tend to experience chronic leakage or seal problems because the particulate matter prevents the ball, disk, or gate from completely closing against the stationary surface.

Simply installing a valve with the same inlet and discharge size as the piping used in the process is not acceptable. In most cases, the valve must be larger than the piping to compensate for flow restrictions within the valve.

Operating Methods

Operating methods for control valves, which are designed to control or direct gas and liquid flow through process systems or fluid-power circuits, range from manual to remote, automatic operation. The key parameters that govern the operation of valves are the speed of the control movement and the impact of speed on the system. This is especially important in process systems.

Hydraulic hammer, or the shock wave generated by the rapid change in the flow rate of liquids within a pipe or vessel, has a serious and negative impact on all components of the process system. For example, instantaneously closing a large flow-control valve may generate in excess of 3 million foot-pounds of force on the entire system upstream of the valve. This shock wave can cause catastrophic failure of upstream valves, pumps, welds, and other system components.

Changes in flow rate, pressure, direction, and other controllable variables must be gradual enough to permit a smooth transition. Abrupt changes in valve position should be avoided. Neither the valve installation nor the control mechanism should permit complete shut off, referred to as *deadheading*, of any circuit in a process system.

Restricted flow forces system components, such as pumps, to operate outside of their performance envelope. This reduces equipment reliability and sets the stage for catastrophic failure or abnormal system performance. In applications in which radical changes in flow are required for normal system operation, control valves should be configured to provide an adequate bypass for surplus flow to protect the system.

For example, systems that must have close control of flow should use two proportioning valves that act in tandem to maintain a balanced hydraulic or aerodynamic system. The primary or master valve should control flow to the downstream process. The second valve, slaved to the master, should divert excess flow to a bypass loop. This master-slave approach ensures that the pumps and other upstream system components are permitted to operate well within their operating envelopes.

FLUID POWER

Fluid power control valves are used on pneumatic and hydraulic systems or circuits.

Configuration

The configuration of fluid power control valves varies with their intended application. The more common configurations include one-way, two-way, three-way, and four-way.

One-Way

One-way valves are typically used for flow and pressure control in fluid-power circuits (Figure 13.9). Flow-control valves regulate the flow of hydraulic fluid or gases in these systems. Pressure-control valves, in the form of regulators or relief valves, control the amount of pressure transmitted downstream from the valve. In most cases, the types of valves used for flow control are smaller versions of the types of valves used in process control. These include ball, gate, globe, and butterfly valves.

Pressure-control valves have a third port to vent excess pressure and prevent it from affecting the downstream piping. The bypass or exhaust port has an internal flow-control device, such as a diaphragm or piston, that opens at predetermined setpoints to permit the excess pressure to bypass the valve's primary discharge. In pneumatic circuits, the bypass port vents to the

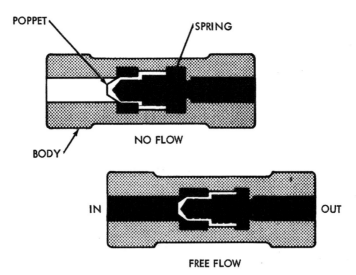

Figure 13.9 One-way fluid-power valve.

atmosphere. In hydraulic circuits, it must be connected to a piping system that returns to the hydraulic reservoir.

Two-Way

A two-way valve has two functional flow-control ports. A two-way, sliding-spool directional control valve is shown in Figure 13.10. As the spool moves back and forth, it either allows fluid to flow through the valve or prevents it from flowing. In the open position, the fluid enters the inlet port, flows around the shaft of the spool, and flows through the outlet port. Because the forces in the cylinder are equal when the valve is open, the spool cannot move back and forth. In the closed position, one of the spool's pistons simply blocks the inlet port, which prevents flow through the valve.

A number of features common to most sliding-spool valves are shown in Figure 13.10. The small ports at either end of the valve housing provide a path for fluid that leaks past the spool to flow to the reservoir. This prevents pressure from building up against the ends of the pistons, which would hinder the movement of the spool. When these valves become worn, they may lose balance because of greater leakage on one side of the spool than on the other. This can cause the spool to stick as it attempts to move back and forth. Therefore, small grooves are machined around the sliding surface of the piston. In hydraulic valves, leaking liquid encircles the piston, keeping the contacting surfaces lubricated and centered.

Three-way

Three-way valves contain a pressure port, cylinder port, and return or exhaust port (Figure 13.11). The three-way directional control valve is designed to operate an actuating unit in one direction. It is returned to its original position either by a spring or the load on the actuating unit.

Four-way

Most actuating devices require system pressure to operate in two directions. The four-way directional control valve, which contains four ports, is used to control

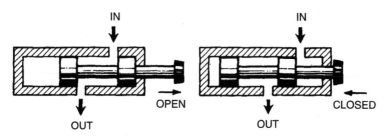

Figure 13.10 Two-way, fluid-power valve.

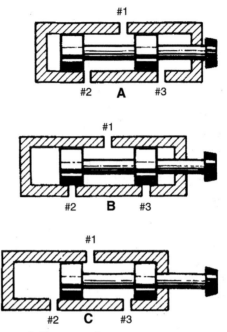

Figure 13.11   Three-way, fluid-power valve.

the operation of such devices (Figure 13.12). The four-way valve also is used in some systems to control the operation of other valves. It is one of the most widely used directional-control valves in fluid-power systems.

The typical four-way directional control valve has four ports: pressure port, return port, and two cylinder or work (output) ports. The pressure port is connected to the main system-pressure line, and the return port is connected to the reservoir return line. The two outputs are connected to the actuating unit.

Performance

The criteria that determine performance of fluid-power valves are similar to those for process-control valves. As with process-control valves, fluid-power valves also must be selected based on their intended application and function.

Installation

When installing fluid power control valves, piping connections are made either directly to the valve body or to a manifold attached to the valve's base. Care

AIR INTRODUCED THROUGH
THIS PASSAGE PUSHES
AGAINST THE PISTON
WHICH SHIFTS THE
SPOOL TO THE RIGHT

CENTERING
WASHERS

SPRINGS PUSH AGAINST
CENTERING WASHERS TO
CENTER THE SPOOL WHEN
NO AIR IS APPLIED

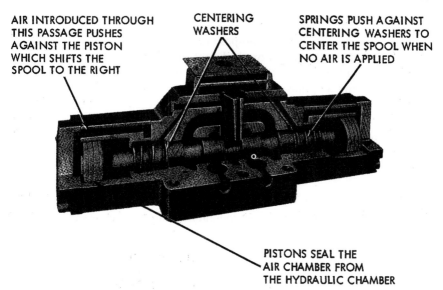

PISTONS SEAL THE
AIR CHAMBER FROM
THE HYDRAULIC CHAMBER

Figure 13.12 Four-way, fluid-power valves.

should be taken to ensure that piping is connected to the proper valve port. The schematic diagram that is affixed to the valve body will indicate the proper piping arrangement as well as the designed operation of the valve. In addition, the ports on most fluid-power valves are generally clearly marked to indicate their intended function.

In hydraulic circuits, the return or common ports should be connected to a return line that directly connects the valve to the reservoir tank. This return line should not need a pressure-control device but should have a check valve to prevent reverse flow of the hydraulic fluid.

Pneumatic circuits may be vented directly to atmosphere. A return line can be used to reduce noise or any adverse effect that locally vented compressed air might have on the area.

Operating Methods

The function and proper operation of a fluid-power valve are relatively simple. Most of these valves have a schematic diagram affixed to the body that clearly explains how to operate the valve.

Valves

Figure 13.13 is a schematic of a two-position, cam-operated valve. The primary actuator, or cam, is positioned on the left of the schematic and any secondary actuators are on the right. In this example, the secondary actuator consists of a spring return and a spring-compensated limit switch. The schematic indicates that when the valve is in the neutral position (right box), flow is directed from the inlet (P) to work port A. When the cam is depressed, the flow momentarily shifts to work port B. The secondary actuator, or spring, automatically returns the valve to its neutral position when the cam returns to its extended position. In these schematics, T indicates the return connection to the reservoir.

Figure 13.14 illustrates a typical schematic of a two-position and three-position directional control valve. The boxes contain flow direction arrows that indicate

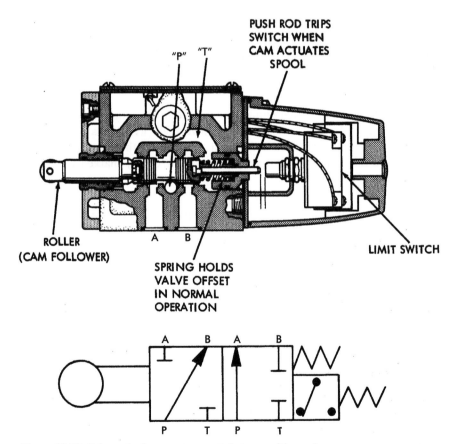

Figure 13.13 Schematic for a cam-operated, two-position valve.

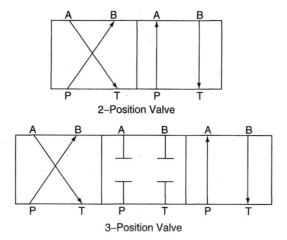

2–Position Valve

3–Position Valve

Figure 13.14 Schematic of two-position and three-position valves.

the flow path in each of the positions. The schematics do not include the actuators used to activate or shift the valves between positions.

In a two-position valve, the flow path is always directed to one of the work ports (A or B). In a three-position valve, a third or neutral position is added. In this figure, a type 2 center position is used. In the neutral position, all ports are blocked and no flow through the valve is possible.

Figure 13.15 is the schematic for the center or neutral position of three-position directional control valves. Special attention should be given to the type of center position that is used in a hydraulic control valve. When type 2, 3, and 6 (see Figure 13.15) are used, the upstream side of the valve must have a relief or bypass valve installed. Since the pressure port is blocked, the valve cannot relieve pressure on the upstream side of the valve. The type 4 center position, called a *motor spool*, permits the full pressure and volume on the upstream side of the valve to flow directly to the return line and storage reservoir. This is the recommended center position for most hydraulic circuits.

The schematic affixed to the valve includes the primary and secondary actuators used to control the valve. Figure 13.16 provides the schematics for three actuator-controlled valves, as follows: (1) double-solenoid, spring-centered, three-position valve; solenoid-operated, spring-return, two-position valve; double-solenoid, detented, two-position valve.

The top schematic represents a double-solenoid, spring-centered, three-position valve. When neither of the two solenoids is energized, the double springs ensure

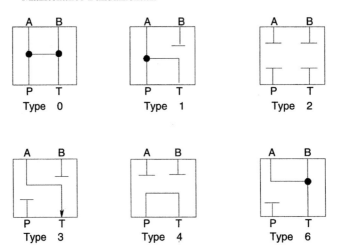

Figure 13.15 Schematic for center or neutral configurations of three-position valves.

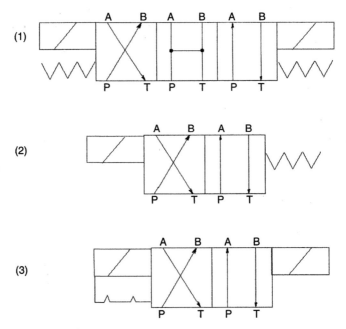

Figure 13.16 Actuator-controlled valve schematics.

that the valve is in its center or neutral position. In this example, a type 0 (see Figure 13.15) configuration is used. This neutral-position configuration equalizes the pressure through the valve. Since the pressure port is open to both work ports and the return line, pressure is equalized throughout the system. When the left or primary solenoid is energized, the valve shifts to the left-hand position and directs pressure to work port B. In this position, fluid in the A-side of the circuit returns to the reservoir. As soon as the solenoid is de-energized, the valve shifts back to the neutral or center position. When the secondary (i.e., right) solenoid is energized, the valve redirects flow to port A, and port B returns fluid to the reservoir.

The middle schematic represents a solenoid-operated, spring-return, two-position valve. Unless the solenoid is energized, the pressure port (P) is connected to work port A. While the solenoid is energized, flow is redirected to work port B. The spring return ensures that the valve is in its neutral (i.e., right) position when the solenoid is de-energized.

The bottom schematic represents a double-solenoid, detented, two-position valve. The solenoids are used to shift the valve between its two positions. A secondary device, called a detent, is used to hold the valve in its last position until the alternate solenoid is energized. Detent configuration varies with the valve type and manufacturer. However, all configurations prevent the valve's control device from moving until a strong force, such as that provided by the solenoid, overcomes its locking force.

Actuators

As with process-control valves, actuators used to control fluid-power valves have a fundamental influence on performance. The actuators must provide positive, real-time response to control inputs. The primary types of actuators used to control fluid-power valves are mechanical, pilot, and solenoid.

Mechanical The use of manually controlled mechanical valves is limited in both pneumatic and hydraulic circuits. Generally, this type of actuator is used only on isolation valves that are activated when the circuit or fluid-power system is shut down for repair or when direct operator input is required to operate one of the system components.

Manual control devices (e.g., levers, cams, or palm buttons) can be used as the primary actuator on most fluid-power control valves. Normally, these actuators are used in conjunction with a secondary actuator, such a spring return or detent, to ensure proper operation of the control valve and its circuit.

Spring returns are used in applications in which the valve is designed to stay open or shut only when the operator holds the manual actuator in a particular position. When the operator releases the manual control, the spring returns the valve to the neutral position.

Valves with a detented secondary actuator are designed to remain in the last position selected by the operator until manually moved to another position. A detent actuator is simply a notched device that locks the valve in one of several pre-selected positions. When the operator applies force to the primary actuator, the valve shifts out of the detent and moves freely until the next detent is reached.

Pilot Although a variety of pilot actuators are used to control fluid-power valves, they all work on the same basic principle. A secondary source of fluid or gas pressure is applied to one side of a sealing device, such as a piston or diaphragm. As long as this secondary pressure remains within pre-selected limits, the sealing device prevents the control valve's flow-control mechanism (i.e., spool or poppet) from moving. However, if the pressure falls outside of the pre-selected window, the actuator shifts and forces the valve's primary mechanism to move to another position.

This type of actuator can be used to sequence the operation of several control valves or operations performed by the fluid-power circuit. For example, a pilot-operated valve is used to sequence the retraction of an airplane's landing gear. The doors that conceal the landing gear when retracted cannot close until the gear is fully retracted. A pilot-operated valve senses the hydraulic pressure in the gear-retraction circuit. When the hydraulic pressure reaches a pre-selected point that indicates that the gear is fully retracted, the pilot-actuated valve triggers the closure circuit for the wheel-well doors.

Solenoid Solenoid valves are widely used as actuators for fluid-power systems. This type of actuator consists of a coil that generates an electrical field when energized. The magnetic forces generated by this field force a plunger that is attached to the main valve's control mechanism to move within the coil. This movement changes the position of the main valve.

In some applications, the mechanical force generated by the solenoid coil is not sufficient to move the main valve's control mechanism. When this occurs, the solenoid actuator is used in conjunction with a pilot actuator. The solenoid coil opens the pilot port, which uses system pressure to shift the main valve.

Solenoid actuators are always used with a secondary actuator to provide positive control of the main valve. Because of heat build up, solenoid actuators must be limited to short-duration activation. A brief burst of electrical energy is transmitted to the solenoid's coil and the actuation triggers a movement of the main valve's control mechanism. As soon as the main valve's position is changed, the energy to the solenoid coil is shut off.

This operating characteristic of solenoid actuators is important. For example, a normally closed valve that uses a solenoid actuation can be open only when the solenoid is energized. As soon as the electrical energy is removed from the solenoid's coil, the valve returns to the closed position. The reverse is true of a normally open valve. The main valve remains open except when the solenoid is energized.

The combination of primary and secondary actuators varies with the specific application. Secondary actuators can be another solenoid or any of the other actuator types that have been previously discussed.

TROUBLESHOOTING

Although there are limited common control valve failure modes, the dominant problems are usually related to leakage, speed of operation, or complete valve failure. Table 13.1 lists the more common causes of these failures.

Special attention should be given to the valve actuator when conducting a Root Cause Failure Analysis. Many of the problems associated with both process and fluid-power control valves are actually actuator problems.

In particular, remotely controlled valves that use pneumatic, hydraulic, or electrical actuators are subject to actuator failure. In many cases, these failures are the reason a valve fails to properly open, close, or seal. Even with manually controlled valves, the true root cause can be traced to an actuator problem. For example, when a manually operated process-control valve is jammed open or closed, it may cause failure of the valve mechanism. This over-torquing of the valve's sealing device may cause damage or failure of the seal, or it may freeze the valve stem. Either of these failure modes results in total valve failure.

Table 13.1 Common Failure Modes of Control Valves

	The Causes	**The Problem**						
		Valve Fails to Open	Valve Fails to Close	Leakage Through Valve	Leakage Around Stem	Excessive Pressure Drop	Opens/Closes Too Fast	Opens/Closes Too Slow
Manually Actuated	Dirt/Debris Trapped in Valve Seat		●	●				
	Excessive Wear		●	●				
	Galling	●	●					
	Line Pressure Too High	●	●	●	●	●		
	Mechanical Damage	●	●					
	Not Packed Properly				●			
	Packed Box Too Loose				●			
	Packing Too Tight	●	●					
	Threads/Lever Damaged	●	●					
	Valve Stem Bound	●	●					
	Valve Undersized					●		●
Pilot Actuated	Dirt/Debris Trapped In Valve Seat	●	●	●				
	Galling	●	●					
	Mechanical Damage (Seals, Seat)	●	●	●				
	Pilot Port Blocked/Plugged	●	●	●				
	Pilot Pressure Too High		●				●	
	Pilot Pressure Too Low	●		●				●
Solenoid Actuated	Corrosion	●	●	●				
	Dirt/Debris Trapped in Valve Seat	●	●	●				
	Galling	●	●					
	Line Pressure Too High	●	●	●	●			●
	Mechanical Damage	●	●	●				
	Solenoid Failure	●	●					
	Solenoid Wiring Defective	●	●					
	Wrong Type of Valve (N-O, N-C)	●	●					

14

CONVEYORS

Conveyors are used to transport materials from one location to another within a plant or facility. The variety of conveyor systems is almost infinite, but the two major classifications used in typical chemical plants are pneumatic and mechanical. Note that the power requirements of a pneumatic conveyor system are much greater than for a mechanical conveyor of equal capacity. However, both systems offer some advantages.

PNEUMATIC

Pneumatic conveyors are used to transport dry, free-flowing, granular material in suspension within a pipe or duct. This is accomplished by the use of a high-velocity air stream or by the energy of expanding compressed air within a comparatively dense column of fluidized or aerated material. Principal uses are (1) dust collection, (2) conveying soft materials such as flake or tow, and (3) conveying hard materials such as fly ash, cement, and sawdust.

The primary advantages of pneumatic conveyor systems are the flexibility of piping configurations and the fact that they greatly reduce the explosion hazard. Pneumatic conveyors can be installed in almost any configuration required to meet the specific application. With the exception of the primary driver, there are no moving parts that can fail or cause injury. However, when they are used to transport explosive materials, there is still some potential for static charge buildup that could cause an explosion.

Configuration

A typical pneumatic conveyor system consists of Schedule-40 pipe or ductwork, which provides the primary flow path used to transport the conveyed material. Motive power is provided by the primary driver, which can be a fan, fluidizer, or positive-displacement compressor.

Performance

Pneumatic conveyor performance is determined by the following factors: (1) primary-driver output, (2) internal surface of the piping or ductwork, and (3) condition of the transported material. Specific factors affecting performance include motive power, friction loss, and flow restrictions.

Motive Power

The motive power is provided by the primary driver, which generates the gas (typically air) velocity required to transport material within a pneumatic conveyor system. Therefore the efficiency of the conveying system depends on the primary driver's operating condition.

Friction Loss

Friction loss within a pneumatic conveyor system is a primary source of efficiency loss. The piping or ductwork must be properly sized to minimize friction without lowering the velocity below the value needed to transport the material.

Flow Restrictions

An inherent weakness of pneumatic conveyor systems is their potential for blockage. The inside surfaces must be clean and free of protrusions or other defects that can restrict or interrupt the flow of material. In addition, when a system is shut down or the velocity drops below the minimum required to keep the transported material suspended, the product will drop out or settle in the piping or ductwork. In most cases, this settled material would compress and lodge in the piping. The restriction caused by this compacted material will reduce flow and eventually result in a complete blockage of the system.

Another major contributor to flow restrictions is blockage caused by system backups. This occurs when the end point of the conveyor system (i.e., storage silo, machine, or vessel) cannot accept the entire delivered flow of material. As the transported material backs up in the conveyor piping, it compresses and forms a solid plug that must be manually removed.

Installation

All piping and ductwork should be as straight and short as possible. Bends should have a radius of at least three diameters of the pipe or ductwork. The diameter should be selected to minimize friction losses and maintain enough velocity to prevent settling of the conveyed material. Branch lines should be configured to match as closely as possible the primary flow direction and avoid 90-degree angles to the main line. The area of the main conveyor line at any point along its run should be 20–25% greater than the sum of all its branch lines.

When vertical runs are short in proportion to the horizontal runs, the size of the riser can be restricted to provide the additional velocity if needed. If the vertical runs are long, the primary or a secondary driver must provide sufficient velocity to transport the material.

Clean-outs, or drop-legs, should be installed at regular intervals throughout the system to permit foreign materials to drop out of the conveyed material. In addition, they provide the means to remove materials that drop out when the system is shut down or air velocity is lost. It is especially important to install adequate clean-out systems near flow restrictions and at the end of the conveyor system.

Operating Methods

Pneumatic conveyor systems must be operated properly to prevent chronic problems, with the primary concern being to maintain constant flow and velocity. If either of these variables is permitted to drop below the system's design envelope, partial or complete blockage of the conveyor system will occur.

Constant velocity can be maintained only when the system is operated within its performance envelope and when regular clean-out is part of the normal operating practice. In addition, the primary driver must be in good operating condition. Any deviation in the primary driver's efficiency reduces the velocity and can result in partial or complete blockage.

The entire pneumatic-conveyor system should be completely evacuated before shutdown to prevent material from settling in the piping or ductwork. In non-continuous applications, the conveyor system should be operated until all material within the conveyor's piping is transported to its final destination. Material that is allowed to settle will compact and partially block the piping. Over time, this will cause a total blockage of the conveyor system.

MECHANICAL

A variety of mechanical conveyor systems are used in chemical plants. These systems generally are composed of chain- or screw-type mechanisms.

Chain

A commonly used chain-type system is a flight conveyor (e.g., Hefler conveyor), which is used to transport granular, lumpy, or pulverized materials along a horizontal or inclined path within totally enclosed ductwork. The Hefler systems generally have lower power requirements than the pneumatic conveyor and have the added benefit of preventing product contamination. This section focuses primarily on the Hefler-type conveyor because it is one of the most commonly used systems.

Configuration

The most common chain conveyor uses a center- or double-chain configuration to provide positive transfer of material within its ductwork. Both chain configurations use hardened bars or U-shaped devices that are an integral part of the chain to drag the conveyed material through the ductwork.

Performance

Data used to determine a chain conveyor's capacity and the size of material that can be conveyed are presented in Table 14.1. Note that these data are for level conveyors. When inclined, capacity data obtained from Table 14.1 must be multiplied by the factors provided in Table 14.2.

Installation

The primary installation concerns with Hefler-type conveyor systems are the ductwork and primary-drive system.

Table 14.1 Approximate Capacities of Chain Conveyors

Flight Width and Depth (Inches)	Quantity of material (Ft³/Ft)	Approximate Capacity (Short Tons/Hour)	Lump Size Single Strand (Inches)	Lump Size Dual Strand (Inches)
12X6	0.40	60	31.5	4.0
15X6	0.49	73	41.5	5.0
18X6	0.56	84	5.0	6.0
24X8	1.16	174		10.0
30X10	1.60	240		14.0
36X12	2.40	360		16.0

Table 14.2 Capacity Correction Factors for Inclined Chain Conveyors

Inclination, degrees	20	25	30	35
Factor	0.9	0.8	0.7	0.6

Ductwork The inside surfaces of the ductwork must be free of defects or protrusions that interfere with the movement of the conveyor's chain or transported product. This is especially true at the joints. The ductwork must be sized to provide adequate chain clearance but should not be large enough to have areas in which the chain drive bypasses the product.

A long horizontal run followed by an upturn is inadvisable because of radial thrust. All bends should have a large radius to permit smooth transition and to prevent material buildup. As with pneumatic conveyors, the ductwork should include clean-out ports at regular intervals for ease of maintenance.

Primary Drive System Most mechanical conveyors use a primary-drive system that consists of an electric motor and a speed-increaser gearbox. See Chapter 8 for more information on gear-drive performance and operation criteria.

The drive-system configuration may vary depending on the specific application or vendor. However, all configurations should include a single-point-of-failure device, such as a shear pin, to protect the conveyor. The shear pin is critical in this type of conveyor because it is prone to catastrophic failure caused by blockage or obstructions that may lock the chain. Use of the proper shear pin prevents major damage from occurring to the conveyor system.

For continuous applications, the primary-drive system must have adequate horsepower to handle a fully loaded conveyor. Horsepower requirements should be determined based on the specific product's density and the conveyor's maximum-capacity rating.

For intermittent applications, the initial startup torque is substantially greater than for a continuous operation. Therefore selection of the drive system and the designed failure point of the shear device must be based on the maximum startup torque of a fully loaded system.

If either the drive system or designed failure point is not properly sized, this type of conveyor is prone to chronic failures. The predominant failures are frequent breakage of the shear device and trips of the motor's circuit breaker caused by excessive startup amp loads.

Operating Methods

Most mechanical conveyors are designed for continuous operation and may exhibit problems in intermittent-service applications. The primary problem is the startup torque for a fully loaded conveyor. This is especially true for conveyor systems handling material that tends to compact or compress on settling in a vessel, such as the conveyor trough.

The only positive method of preventing excessive startup torque is to ensure that the conveyor is completely empty before shutdown. In most cases, this can be accomplished by isolating the conveyor from its supply for a few minutes prior to shutdown. This time delay permits the conveyor to deliver its entire load of product before it is shut off.

In applications in which it is impossible to completely evacuate the conveyor prior to shutdown, the only viable option is to jog, or step start, the conveyor. Step starting reduces the amp load on the motor and should control the torque to prevent the shear pin from failing.

If, instead of step starting, the operator applies full motor load to a stationary, fully loaded conveyor, one of two things will occur: (1) the drive motor's circuit breaker will trip as a result of excessive amp load; or (2) the shear pin installed to protect the conveyor will fail. Either of these failures adversely affects production.

Screw

The screw, or spiral, conveyor is widely used for pulverized or granular, non-corrosive, non-abrasive materials in systems requiring moderate capacities, distances not more than about 200 feet, and moderate inclines ($\leq$ 35 degrees). It usually costs substantially less than any other type of conveyor and can be made dust-tight by installing a simple cover plate.

Abrasive or corrosive materials can be handled with suitable construction of the helix and trough. Conveyors using special materials, such as hard-faced cast iron and linings or coatings on the components that come into contact with the materials, can be specified in these applications. The screw conveyor will handle lumpy material if the lumps are not large in proportion to the diameter of the screw's helix.

Screw conveyors may be inclined. A standard-pitch helix will handle material on inclines up to 35 degrees. Capacity is reduced in inclined applications, and Table 14.3 provides the approximate reduction in capacity for various inclines.

Table 14.3 Screw Conveyor Capacity Reductions for Inclined Applications

Inclination, degrees	10	15	20	25	30	35
Reduction in capacity, %	10	26	45	58	70	78

Configuration

Screw conveyors have a variety of configurations. Each is designed for specific applications and/or materials. Standard conveyors have a galvanized-steel rotor, or helix, and trough. For abrasive and corrosive materials (e.g., wet ash), both the helix and trough may be hard-faced cast iron. For abrasives, the outer edge of the helix may be faced with a renewable strip of Stellite (a cobalt alloy produced by Haynes Stellite Co.) or other similarly hard material. Aluminum, bronze, Monel, or stainless steel also may be used to construct the rotor and trough.

Short-Pitch Screw

The standard helix used for screw conveyors has a pitch approximately equal to its outside diameter. The short-pitch screw is designed for applications with inclines greater than 29 degrees.

Variable-Pitch Screw

Variable-pitch screws having the short pitch at the feed end automatically control the flow to the conveyor and correctly proportion the load down the screw's length. Screws having what is referred to as a "short section," which has either a shorter pitch or smaller diameter, are self-loading and do not require a feeder.

Cut Flight

Cut-flight conveyors are used for conveying and mixing cereals, grains, and other light material. They are similar to normal flight or screw conveyors and the only difference is the configuration of the paddles or screw. Notches are cut in the flights to improve the mixing and conveying efficiency when handling light, dry materials.

Ribbon Screw

Ribbon screws are used for wet and sticky materials such as molasses, hot tar, and asphalt. This type of screw prevents the materials from building up and altering the natural frequency of the screw. A buildup can cause resonance problems and possibly catastrophic failure of the unit.

Paddle Screw

The paddle-screw conveyor is used primarily for mixing materials such as mortar and paving mixtures. An example of a typical application is churning ashes and water to eliminate dust.

Performance

Process parameters such as density, viscosity, and temperature must be constantly maintained within the conveyor's design operating envelope. Slight variations can affect performance and reliability. In intermittent applications, extreme care should be taken to fully evacuate the conveyor prior to shutdown. In addition, caution must be exercised when re-starting a conveyor in case an improper shutdown was performed and material was allowed to settle.

Power Requirements

The horsepower requirement for the conveyor-head shaft, H, for horizontal screw conveyors can be determined from the following equation:

$$H = (ALN + CWLF) \times 10 - 6$$

where

A = Factor for size of conveyor (see Table 14.4)
C = Material volume, $ft.^3/h$
F = Material factor, unitless (see Table 14.5)
L = Length of conveyor, ft.
N = Conveyor rotation speed (rpm)
W = Density of material, $lb/ft.^3$

In addition to H, the motor size depends on the drive efficiency (E) and a unitless allowance factor (G), which is a function of H. Values for G are found in Table 14.6. The value for E is usually 90%.

$$\text{Motor Hp} = HG/E$$

Table 14.4 Factor A for Self-Lubricating Bronze Bearings

Conveyor Diameter, in.	6	9	10	12	14	16	18	20	24
Factor A	54	96	114	171	255	336	414	510	690

Table 14.5 Power Requirements by Material Group

Material Group	Max. Cross-section % Occupied by the Material	Max. Density of Material, lb/ft^3	Max. rpm for 6-inch diameter	Max. rpm for 20-inch diameter
1	45	50	170	110
2	38	50	120	75
3	31	75	90	60
4	25	100	70	50
5	12½		30	25

Group 1   F factor is 0.5 for light materials such as barley, beans, brewers' grains (dry), coal (pulverized), corn meal, cottonseed meal, flaxseed, flour, malt, oats, rice, and wheat.

Group 2   Includes fines and granular material. The values of F are: alum (pulverized), 0.6; coal (slack or fines), 0.9; coffee beans, 0.4; sawdust, 0.7; soda ash (light), 0.7; soybeans, 0.5; fly ash, 0.4.

Group 3   Includes materials with small lumps mixed with fines. Values of F are alum, 1.4; ashes (dry), 4.0; borax, 0.7; brewers' grains (wet), 0.6; cottonseed, 0.9; salt, coarse or fine, 1.2; soda ash (heavy), 0.7.

Group 4   Includes semi-abrasive materials, fines, granular, and small lumps. Values of F are: acid phosphate (dry), 1.4; bauxite (dry), 1.8; cement (dry), 1.4; clay, 2.0; Fuller's earth, 2.0; lead salts, 1.0; limestone screenings, 2.0; sugar (raw), 1.0; white lead, 1.0; sulfur (lumpy), 0.8; zinc oxide, 1.0.

Group 5   Includes abrasive lumpy materials, which must be kept from contact with hanger bearings. Values of F are: wet ashes, 5.0; flue dirt, 4.0; quartz (pulverized), 2.5; silica sand, 2.0; sewage sludge (wet and sandy), 6.0.

Table 14.5 gives the information needed to estimate the power requirement: percentages of helix loading for five groups of material, maximum material density or capacity, allowable speeds for 6-inch and 20-inch diameter screws, and the factor F.

Table 14.6 Allowance Factor

H, Hp	1	1–2	2–4	4–5	5
G	2	1.5	1.25	1.1	1.0

Volumetric Efficiency

Screw-conveyor performance is also determined by the volumetric efficiency of the system. This efficiency is determined by the amount of slip or bypass generated by the conveyor. The amount of slip in a screw conveyor is primarily determined by three factors: product properties, screw efficiency, and clearance between the screw and the conveyor barrel or housing.

Product Properties

Not all materials or products have the same flow characteristics. Some have plastic characteristics and flow easily. Others do not self-adhere and tend to separate when pumped or mechanically conveyed. As a result, the volumetric efficiency is directly affected by the properties of each product. This also affects screw performance.

Screw Efficiency

Each of the common screw configurations (i.e., short-pitch, variable-pitch, cut flights, ribbon, and paddle) has varying volumetric efficiencies, depending on the type of product that is conveyed. Screw designs or configurations must be carefully matched to the product to be handled by the system.

For most medium- to high-density products in a chemical plant, the variable-pitch design normally provides the highest volumetric efficiency and lowest required horsepower. Cut-flight conveyors are highly efficient for light, non-adhering products such as cereals but are inefficient when handling heavy, cohesive products. Ribbon conveyors are used to convey heavy liquids such as molasses but are not very efficient and have a high slip ratio.

Clearance

Improper clearance is the source of many volumetric efficiency problems. It is important to maintain proper clearance between the outer ring, or diameter, of the screw and the conveyor's barrel, or housing, throughout the operating life of the conveyor. Periodic adjustments to compensate for wear, variations in product, and changes in temperature are essential. While the recommended clearance varies with specific conveyor design and the product to be conveyed, excessive clearance severely affects conveyor performance as well.

Installation

Installation requirements vary greatly with screw-conveyor design. The vendor's Operating and Maintenance (O&M) manuals should be consulted and followed to ensure proper installation. However, as with practically all mechanical equipment, there are basic installation requirements common to all screw conveyors. Installation requirements presented here should be evaluated in conjunction with the vendor's O&M manual. If the information provided here conflicts with the vendor-supplied information, the O&M manual's recommendations should always be followed.

Foundation

The conveyor and its support structure must be installed on a rigid foundation that absorbs the torsional energy generated by the rotating screws. Because of the total overall length of most screw conveyors, a single foundation that supports the entire length and width should be used. There must be enough lateral (i.e., width) stiffness to prevent flexing during normal operation. Mounting conveyor systems on decking or suspended-concrete flooring should provide adequate support.

Support Structure

Most screw conveyors are mounted above the foundation level on a support structure that generally has a slight downward slope from the feed end to the discharge end. While this improves the operating efficiency of the conveyor, it also may cause premature wear of the conveyor and its components.

The support's structural members (i.e., I-beams and channels) must be adequately rigid to prevent conveyor flexing or distortion during normal operation. Design, sizing, and installation of the support structure must guarantee rigid support over the full operating range of the conveyor. When evaluating the structural requirements, variations in product type, density, and operating temperature also must be considered. Since these variables directly affect the torsional energy generated by the conveyor, the worst-case scenario should be used to design the conveyor's support structure.

Product-Feed System

One of the major limiting factors of screw conveyors is their ability to provide a continuous supply of incoming product. While some conveyor designs, such as those having a variable-pitch screw, provide the ability to self-feed, their installation should include a means of ensuring a constant, consistent incoming supply of product.

In addition, the product-feed system must prevent entrainment of contaminates in the incoming product. Normally, this requires an enclosure that seals the product from outside contaminants.

Operating Methods

As previously discussed, screw conveyors are sensitive to variations in incoming product properties and the operating environment. Therefore, the primary operating concern is to maintain a uniform operating envelope at all times, in particular by controlling variations in incoming product and operating environment.

Incoming-Product Variations

Any measurable change in the properties of the incoming product directly affects the performance of a screw conveyor. Therefore the operating practices should limit variations in product density, temperature, and viscosity. If they occur, the conveyor's speed should be adjusted to compensate for them.

For property changes directly related to product temperature, preheaters or coolers can be used in the incoming-feed hopper, and heating/cooling traces can be used on the conveyor's barrel. These systems provide a means of achieving optimum conveyor performance despite variations in incoming product.

Operating-Environment Variations

Changes in the ambient conditions surrounding the conveyor system may also cause deviations in performance. A controlled environment will substantially improve the conveyor's efficiency and overall performance. Therefore, operating practices should include ways to adjust conveyor speed and output to compensate for variations. The conveyor should be protected from wind chill, radical variations in temperature and humidity, and any other environment-related variables.

15

FANS, BLOWERS, AND FLUIDIZERS

Technically, fans and blowers are two separate types of devices that have a similar function. However, the terms are often used interchangeably to mean any device that delivers a quantity of air or gas at a desired pressure. Differences between these two devices are their rotating elements and their discharge-pressure capabilities. Fluidizers are identical to single-stage, screw-type compressors or blowers.

CENTRIFUGAL FANS

The centrifugal fan is one of the most common machines used in industry. It utilizes a rotating element with blades, vanes, or propellers to extract or deliver a specific volume of air or gas. The rotating element is mounted on a rotating shaft that must provide the energy required to overcome inertia, friction, and other factors that restrict or resist air or gas flow in the application. They are generally low-pressure machines designed to overcome friction and either suction or discharge-system pressure.

Configuration

The type of rotating element or wheel that is used to move the air or gas can classify the centrifugal fan. The major classifications are propeller and axial. Axial fans also can be further differentiated by the blade configurations.

Propeller

This type of fan consists of a propeller, or paddle wheel, mounted on a rotating shaft within a ring, panel, or cage. The most widely used propeller fans are found in light- or medium-duty functions such as ventilation units in which air can be moved in any direction. These fans are commonly used in wall mountings to inject air into, or exhaust air from, a space. Figure 15.1 illustrates a belt-driven propeller fan appropriate for medium-duty applications.

This type of fan has a limited ability to boost pressure. Its use should be limited to applications in which the total resistance to flow is less than 1 inch of water. In addition, it should not be used in corrosive environments or where explosive gases are present.

Axial

Axial fans are essentially propeller fans that are enclosed within a cylindrical housing or shroud. They can be mounted inside ductwork or a vessel housing to inject or exhaust air or gas. These fans have an internal motor mounted on spokes or struts to centralize the unit within the housing. Electrical connections and grease fittings are mounted externally on the housing. Arrow indicators on the housing show the direction of airflow and rotation of the shaft, which enables the unit to be correctly installed in the ductwork. Figure 15.2 illustrates an inlet end of a direct-connected, tube-axial fan.

This type of fan should not be used in corrosive or explosive environments, because the motor and bearings cannot be protected. Applications in which concentrations of airborne abrasives are present should also be avoided.

Figure 15.1 Belt-driven propeller fan for medium-duty applications.

Figure 15.2 Inlet end of a direct-connected tube-axial fan.

Axial fans use three primary types of blades or vanes: backward-curved, forward-curved, and radial. Each type has specific advantages and disadvantages.

Backward-Curved Blades The backward-curved blade provides the highest efficiency and lowest sound level of all axial-type centrifugal fan blades. Advantages include the following:

- Moderate to high volumes
- Static pressure range up to approximately 30 inches of water (gauge)
- Highest efficiency of any type of fan
- Lowest noise level of any fan for the same pressure and volumetric requirements
- Self-limiting brake horsepower (BHP) characteristics (Motors can be selected to prevent overload at any volume, and the BHP curve rises to a peak and then declines as volume increases)

The limitations of backward-curved blades are as follows:

- Weighs more and occupies considerably more space than other designs of equal volume and pressure
- Large wheel width
- Not to be used in dusty environments or where sticky or stringy materials are used, because residues adhering to the blade surface cause imbalance and eventual bearing failure

Forward-Curved Blades This design is commonly referred to as a *squirrel-cage* fan. The unit has a wheel with a large number of wide, shallow blades; a very large intake area relative to the wheel diameter; and a relatively slow operational speed. The advantages of forward-curved blades include the following:

- Excellent for any volume at low to moderate static pressure when using clean air
- Occupies approximately the same space as backward-curved blade fan
- More efficient and much quieter during operation than propeller fans for static pressures above approximately 1 inch of water (gauge)

The limitations of forward-curved blades include the following:

- Not as efficient as backward-curved blade fans
- Should not be used in dusty environments or handle sticky or stringy materials that could adhere to the blade surface
- BHP increases as this fan approaches maximum volume, as opposed to backward-curved blade centrifugal fans, which experience a decrease in BHP as they approach maximum volume

Radial Blades Industrial exhaust fans fall into this category. The design is rugged and may be belt-driven or directly driven by a motor. The blade shape varies considerably from flat surfaces to various bent configurations to increase efficiency slightly or to suit particular applications. The advantages of radial-blade fans include the following:

- Best suited for severe duty, especially when fitted with flat radial blades
- Simple construction that lends itself to easy field maintenance
- Highly versatile industrial fan that can be used in extremely dusty environments as well as with clean air
- Appropriate for high-temperature service
- Handles corrosive or abrasive materials

The limitations of radial-blade fans include the following:

- Lowest efficiency in centrifugal-fan group
- Highest sound level in centrifugal-fan group
- BHP increases as fan approaches maximum volume

Performance

A fan is inherently a constant-volume machine. It operates at the same volumetric flow rate (i.e., cubic feet per minute) when operating in a fixed system at a constant speed, regardless of changes in air density. However, the pressure developed and the horsepower required vary directly with the air density.

The following factors affect centrifugal-fan performance: brake horsepower, fan capacity, fan rating, outlet velocity, static efficiency, static pressure, tip speed, mechanical efficiency, total pressure, velocity pressure, natural frequency, and suction conditions. Some of these factors are used in the mathematical relationships that are referred to as Fan Laws.

Brake Horsepower

Brake horsepower (BHP) is the power input required by the fan shaft to produce the required volumetric flow rate (cfm) and pressure.

Fan Capacity

The fan capacity (FC) is the volume of air moved per minute by the fan (cfm). Note: the density of air is 0.075 pounds per cubic foot at atmospheric pressure and 68°F.

Fan Rating

The fan rating predicts the fan's performance at one operating condition, which includes the fan size, speed, capacity, pressure, and horsepower.

Outlet Velocity

The outlet velocity (OV, feet per minute) is the number of cubic feet of gas moved by the fan per minute divided by the inside area of the fan outlet, or discharge area, in square feet.

Static Efficiency

Static efficiency (SE) is not the true mechanical efficiency but is convenient to use in comparing fans. This is calculated by the following equation:

$$\text{Static Eficiency (SE)} = \frac{0.000157 \times \text{FC} \times \text{SP}}{\text{BHP}}$$

Static Pressure

Static pressure (SP) generated by the fan can exist whether the air is in motion or is trapped in a confined space. SP is always expressed in inches of water (gauge).

Tip Speed

The tip speed (TS) is the peripheral speed of the fan wheel in feet per minute (fpm).

$$\text{Tip Speed} = \text{Rotor Diameter} \times \pi \times \text{RPM}$$

Mechanical Efficiency

True mechanical efficiency (ME) is equal to the total input power divided by the total output power.

Total Pressure

Total pressure (TP), inches of water (gauge), is the sum of the velocity pressure and static pressure.

Velocity Pressure

Velocity pressure (VP) is produced by the fan when the air is moving. Air having a velocity of 4,000 fpm exerts a pressure of 1 inch of water (gauge) on a stationary object in its flow path.

Natural Frequency

General-purpose fans are designed to operate below their first natural frequency. In most cases, the fan vendor will design the rotor-support system so that the rotating element's first critical is between 10% and 15% above the rated running speed. While this practice is questionable, it is acceptable if the design speed and rotating-element mass are maintained. However, if either of these two factors changes, there is a high probability that serious damage or premature failure will result.

Inlet-Air Conditions

As with centrifugal pumps, fans require stable inlet conditions. Ductwork should be configured to ensure an adequate volume of clean air or gas, stable inlet pressure, and laminar flow. If the supply air is extracted from the environment, it is subject to variations in moisture, dirt content, barometric pressure, and density. However, these variables should be controlled as much as possible. As a minimum, inlet filters should be installed to minimize the amount of dirt and moisture that enters the fan.

Excessive moisture and particulates have an extremely negative impact on fan performance and cause two major problems: abrasion or tip wear and plate-out. High concentrations of particulate matter in the inlet air act as abrasives that accelerate fan-rotor wear. In most cases, however, this wear is restricted to the high-velocity areas of the rotor, such as the vane or blade tips, but can affect the entire assembly.

Plate-out is a much more serious problem. The combination of particulates and moisture can form "glue" that binds to the rotor assembly. As this contamination builds up on the rotor, the assembly's mass increases, which reduces its natural frequency. If enough plate-out occurs, the fan's rotational speed may coincide with the rotor's reduced natural frequency. With a strong energy source like the running speed, the excitation of the rotor's natural frequency can result in catastrophic fan failure. Even if catastrophic failure does not occur, the vibration energy generated by the fan may cause bearing damage.

Fan Laws

The mathematical relationships referred to as *Fan Laws* can be useful when applied to fans operating in a fixed system or to geometrically similar fans.

However, caution should be exercised when using these relationships. They apply only to identical fans and applications. The basic laws are as follows:

- Volume in cubic feet per minute (cfm) varies directly with the rotating speed (rpm)
- Static pressure varies with the rotating speed squared (rpm^2)
- BHP varies with the speed cubed (rpm^3)

The fan-performance curves shown in Figures 15.3 and 15.4 show the performance of the same fan type, but designed for different volumetric-flow rates, operating in the same duct system handling air at the same density.

Curve #1 is for a fan designed to handle 10,000 cfm in a duct system whose calculated system resistance is determined to be 1 inch of water (gauge). This fan

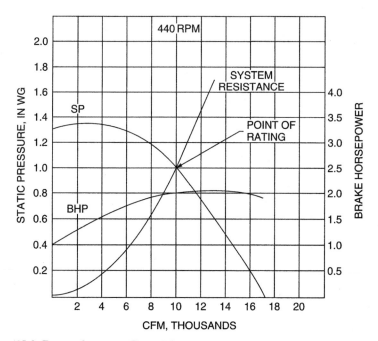

Figure 15.3 Fan-performance Curve #1.

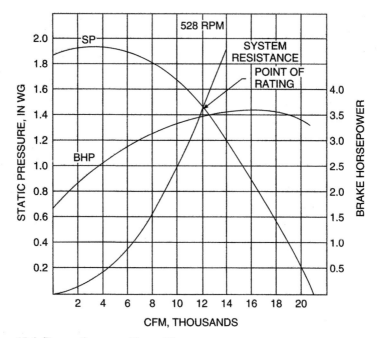

Figure 15.4 Fan-performance Curve #2.

will operate at the point where the fan pressure (SP) curve intersects the system resistance curve (TSH). This intersection point is called the Point of Rating. The fan will operate at this point provided the fan's speed remains constant and the system's resistance does not change. The system-resistance curve illustrates that the resistance varies as the square of the volumetric flow rate (cfm). The BHP of the fan required for this application is 2 Hp.

Curve #2 illustrates the situation if the fan's design capacity is increased by 20%, increasing output from 10,000 to 12,000 cfm. Applying the Fan Laws, the calculations are:

$$\text{New rpm} = 1.2 \times 440$$
$$= 528 \text{ rpm } (20\% \text{ increase})$$
$$\text{New SP} = 1.2 \times 1.2 \times 1 \text{ inch water (gauge)}$$
$$= 1.44 \text{ inches } (44\% \text{ increase})$$
$$\text{New TSH} = \text{New SP} = 1.44 \text{ inches}$$
$$\text{New BHP} = 1.2 \times 1.2 \times 1.2 \times 2$$
$$= 1.73 \times 2$$
$$= 3.46 \text{ Hp } (73\% \text{ increase})$$

The curve representing the system resistance is the same in both cases, since the system has not changed. The fan will operate at the same relative point of rating and will move the increased volume through the system. The mechanical and static efficiencies are unchanged.

The increased BHP required to drive the fan is a very important point to note. If a 2-Hp motor had driven the Curve #1 fan, the Curve #2 fan needs a 3.5-hp motor to meet its volumetric requirement.

Centrifugal fan selection is based on rating values such as air flow, rpm, air density, and cost. Table 15.1 is a typical rating table for a centrifugal fan. Table 15.2 provides air-density ratios.

Installation

Proper fan installation is critical to reliable operation. Suitable foundations, adequate bearing-support structures, properly sized ductwork, and flow-control devices are the primary considerations.

Foundations

As with any other rotating machine, fans require a rigid, stable foundation. With the exception of inline fans, they must have a concrete footing or pad that is properly sized to provide a stable footprint and prevent flexing of the rotor-support system.

Bearing-Support Structures

In most cases, with the exception of inline configurations, fans are supplied with a vendor-fabricated base. Bases normally consist of fabricated metal stands that support the motor and fan housing. The problem with most of the fabricated bases is that they lack the rigidity and stiffness to prevent flexing or distortion of the fan's rotating element. The typical support structure is composed of relatively light-gauge material (3/16 in.) and does not have the cross-bracing or structural stiffeners needed to prevent distortion of the rotor assembly. Because of this limitation, many plants fill the support structure with concrete or other solid material.

However, this approach does little to correct the problem. When the concrete solidifies, it pulls away from the sides of the support structure. Without direct bonding and full contact with the walls of the support structure, stiffness is not significantly improved.

The best solution to this problem is to add cross-braces and structural stiffeners. If they are properly sized and affixed to the support structure, the stiffness can be improved and rotor distortion reduced.

Table 15.1 Typical Rating Table for a Centrifugal Fan

CFM	OV	1/4" SP		3/8" SP		1/2" SP		5/8" SP		3/4" SP	
		RPM	BHP	RPM	BHP	RPM	BHP	RPM	BHP	RPM	BHP
7456	800	262	0.45	289	0.60	314	0.75	337	0.92	360	1.09
8388	900	281	0.55	306	0.72	330	0.89	351	1.06	372	1.25
9320	1000	199	0.66	325	0.85	347	1.04	368	1.23	387	1.43
10252	1100	319	0.79	343	1.00	365	1.21	385	1.42	403	1.63
11184	1200	338	0.93	362	1.17	383	1.40	402	1.63	420	1.85
12116	1300	358	1.10	381	1.35	402	1.61	421	1.85	438	2.10
13048	1400	379	1.29	401	1.56	421	1.83	439	2.10	456	2.37
13980	1500	401	1.50	420	1.78	440	2.08	458	2.37	475	2.66
14912	1600	422	1.74	441	2.03	459	2.35	477	2.67	494	2.98
15844	1700	444	2.01	462	2.32	479	2.65	496	2.98	513	3.32
16776	1800	467	2.31	483	2.63	499	2.97	516	3.33	532	3.68
17708	1900	489	2.65	504	2.98	520	3.33	536	3.70	551	4.07
18640	2000	512	3.02	526	3.36	541	3.72	556	4.10	571	4.49
19572	2100	535	3.43	548	3.77	562	4.15	576	4.53	590	4.95
20504	2200	558	3.87	570	4.23	584	4.61	597	5.02	610	5.43
21436	2300	582	4.36	593	4.72	605	5.12	618	5.54	631	5.95
22368	2400	605	4.89	616	5.26	627	5.67	640	6.10	652	6.54
23300	2500	628	5.46	639	5.85	650	6.26	661	6.70	673	7.16
24232	2600	652	6.09	662	6.48	672	6.90	683	7.34	694	7.81
25164	2700	676	6.75	685	7.15	695	7.58	705	8.04	715	8.52
26096	2800	700	7.47	708	7.86	718	8.32	727	8.78	738	9.27
27028	2900	723	8.24	732	8.66	741	9.11	750	9.58	760	10.08

7/8" SP		1" SP		1 1/4" SP		1 1/2" SP		1 3/4" SP	
RPM	BHP	RPM	BHP	RPM	BHP	RPM	BHP	RPM	BHP
382	1.27	403	1.46	444	1.85	483	2.28	520	2.73
393	1.44	413	1.63	451	2.05	488	2.49	523	2.96
406	1.63	425	1.83	461	2.27	495	2.72	529	3.21
421	1.84	439	2.06	473	2.51	505	2.99	537	3.50
438	2.08	454	2.31	486	2.79	517	3.29	547	3.81
455	2.34	471	2.59	501	3.09	531	3.62	559	4.16
473	2.63	488	2.90	517	3.43	545	3.98	572	4.54
491	2.94	506	3.23	534	3.79	561	4.37	587	4.96
509	3.28	524	3.58	552	4.19	578	4.79	603	5.41
528	3.64	542	3.97	570	4.61	595	5.25	619	5.90
547	4.03	561	4.38	588	5.06	613	5.74	637	6.42
566	4.45	580	4.81	606	5.54	631	6.26	654	6.98
585	4.89	599	5.28	625	6.05	649	6.81	672	7.57
604	5.36	618	5.78	644	6.59	668	7.40	690	8.19
624	5.87	637	6.30	663	7.16	686	8.01	708	8.85
644	6.41	657	6.86	682	7.77	705	8.66	727	9.54
664	6.99	677	7.46	701	8.41	724	9.35	746	10.27
685	7.63	697	8.10	721	9.09	743	10.07	765	11.04
706	8.30	717	8.77	740	9.80	762	10.83	784	11.84
727	9.01	738	9.53	760	10.56	782	11.63	803	12.69
748	9.78	759	10.30	780	11.35	801	12.46	822	13.57
770	10.60	780	11.13	800	12.20	821	13.35	841	14.49

Table 15.2 Air-Density Ratios

Air temp., °F	Altitude, ft. above sea level					
	0	1,000	2,000	3,000	4,000	5,000
	Barometric pressure, in. mercury					
	29.92	28.86	27.82	26.82	25.84	24.90
70	1.000	0.964	0.930	0.896	0.864	0.832
100	0.946	0.912	0.880	0.848	0.818	0.787
150	0.869	0.838	0.808	0.770	0.751	0.723
200	0.803	0.774	0.747	0.720	0.694	0.668
250	0.747	0.720	0.694	0.669	0.645	0.622
300	0.697	0.672	0.648	0.624	0.604	0.580
350	0.654	0.631	0.608	0.586	0.565	0.544
400	0.616	0.594	0.573	0.552	0.532	0.513
450	0.582	0.561	0.542	0.522	0.503	0.484
500	0.552	0.532	0.513	0.495	0.477	0.459
550	0.525	0.506	0.488	0.470	0.454	0.437
600	0.500	0.482	0.465	0.448	0.432	0.416
650	0.477	0.460	0.444	0.427	0.412	0.397
700	0.457	0.441	0.425	0.410	0.395	0.380

Air temp., °F	Altitude, ft. above sea level						
	6,000	7,000	8,000	9,000	10,000	15,000	20,000
	Barometric pressure, in. mercury						
	23.98	23.09	22.22	21.39	20.58	16.89	13.75
70	0.801	0.772	0.743	0.714	0.688	0.564	0.460
100	0.758	0.730	0.703	0.676	0.651	0.534	0.435
150	0.696	0.671	0.646	0.620	0.598	0.490	0.400
200	0.643	0.620	0.596	0.573	0.552	0.453	0.369
250	0.598	0.576	0.555	0.533	0.514	0.421	0.344
300	0.558	0.538	0.518	0.498	0.480	0.393	0.321
350	0.524	0.505	0.486	0.467	0.450	0.369	0.301
400	0.493	0.476	0.458	0.440	0.424	0.347	0.283
450	0.466	0.449	0.433	0.416	0.401	0.328	0.268
500	0.442	0.426	0.410	0.394	0.380	0.311	0.254
550	0.421	0.405	0.390	0.375	0.361	0.296	0.242
600	0.400	0.386	0.372	0.352	0.344	0.282	0.230
650	0.382	0.368	0.354	0.341	0.328	0.269	0.219
700	0.366	0.353	0.340	0.326	0.315	0.258	0.210

Ductwork

Ductwork should be sized to provide minimum friction loss throughout the system. Bends, junctions with other ductwork, and any change of direction should provide a clean, direct flow path. All ductwork should be airtight and leak-free to ensure proper operation.

Flow-Control Devices

Fans should always have inlet and outlet dampers or other flow-control devices such as variable-inlet vanes. Without them, it is extremely difficult to match fan performance to actual application demand. The reason for this difficulty is that there are a number of variables (e.g., usage, humidity, and temperature) directly affecting the input-output demands for each fan application. Flow-control devices provide the means to adjust fan operation for actual conditions. Figure 15.5 shows an outlet damper with streamlined blades and linkage arranged to move adjacent blades in opposite directions for even throttling.

Figure 15.5 Outlet damper with streamlined blades and linkage arranged to move adjacent blades in opposite directions for even throttling.

Airflow controllers must be inspected frequently to ensure that they are fully operable and operate in unison with each other. They also must close tightly. Ensure that the control indicators show the precise position of the vanes in all operational conditions. The "open" and "closed" positions should be permanently marked and visible at all times. Periodic lubrication of linkages is required.

Turnbuckle screws on the linkages for adjusting flow rates should never be moved without first measuring the distance between the setpoint markers on each screw. This is important if the adjustments do not produce the desired effect and you wish to return to the original settings.

Operating Methods

Because fans are designed for stable, steady-state operation, variations in speed or load may have an adverse effect on their operating dynamics. The primary operating method that should be understood is output control. Two methods can be used to control fan output: dampers and fan speed.

Dampers

Dampers can be used to control the output of centrifugal fans within the effective control limits. Centrifugal fans have a finite acceptable control range, typically about 15% below and 15% above their design points. Control variations outside of this range severely affect the reliability and useful life of these fans.

The recommended practice is to use an inlet damper rather than a discharge damper for this control function whenever possible. Restricting the inlet with suction dampers can effectively control the fan's output. When using dampers to control fan performance, however, caution should be exercised to ensure that any changes remain within the fan's effective control range.

Fan Speed

Varying fan speed is an effective means of controlling a fan's performance. As defined by the Fan Laws, changing the rotating speed of the fan can directly control both volume and pressure. However, caution must be used when changing fan speed. All rotating elements, including fan rotors, have one or more critical speeds. When the fan's speed coincides with one of the critical speeds, the rotor assembly becomes extremely unstable and could fail catastrophically.

In most general-purpose applications, fans are designed to operate between 10% and 15% below their first critical speed. If speed is increased on these fans, there is a good potential for a critical-speed problem. Other applications have fans that are designed to operate between their first and second critical speeds. In this instance, any change up or down may cause the speed to coincide with one of the critical speeds.

BLOWERS

A blower uses mating helical lobes or screws and is used for the same purpose as a fan. Blowers are normally moderate- to high-pressure devices. Blowers are almost identical both physically and functionally to positive-displacement compressors.

FLUIDIZERS

Fluidizers are identical to single-stage, screw-type compressors or blowers. They are designed to provide moderate- to high-pressure transfer of non-abrasive, dry materials.

TROUBLESHOOTING FANS, BLOWERS, AND FLUIDIZERS

Tables 15.3 and 15.4 list the common failure modes for fans, blowers, and fluidizers. Typical problems with these devices include (1) output below rating, (2) vibration and noise, and (3) overloaded driver bearings.

CENTRIFUGAL FANS

Centrifugal fans are extremely sensitive to variations in either suction or discharge conditions. In addition to variations in ambient conditions (i.e., temperature, humidity, etc.), control variables can have a direct effect on fan performance and reliability.

Most of the problems that limit fan performance and reliability are either directly or indirectly caused by improper application, installation, operation, or maintenance. However, the majority are caused by misapplication or poor operating practices. Table 15.4 lists failure modes of centrifugal fans and their causes. Some of the more common failures are aerodynamic instability, plate-out, speed changes, and lateral flexibility.

Aerodynamic Instability

Generally, the control range of centrifugal fans is about 15% above and 15% below their best efficiency point (BEP). When fans are operated outside of this range, they tend to become progressively unstable, which causes the fan's rotor

Table 15.3 Common Failure Modes of Centrifugal Fans

THE CAUSES	Insufficient Discharge Pressure	Intermittent Operation	Insufficient Capacity	Overheated Bearings	Short Bearing Life	Overload on Driver	High Vibration	High Noise Levels	Power Demand Excessive	Motor Trips
Abnormal End Thrust				●			●			
Aerodynamic Instability		●	●	●	●		●	●		
Air Leaks in System	●	●	●							
Bearings Improperly Lubricated						●	●	●		●
Bent Shaft				●	●	●	●		●	
Broken or Loose Bolts or Setscrews				●			●			
Damaged Motor							●			
Damaged Wheel	●		●	●						
Dampers or Variable-inlet Not Properly Adjusted	●		●							
Dirt in Bearings				●			●			
Excessive Belt Tension				●			●			●
External Radiated Heat				●						
Fan Delivering More Than Rated Capacity						●	●			
Fan Wheel or Driver Imbalanced				●			●			
Foreign Material in Fan Causing Imbalance (Plate-out)				●			●	●		
Incorrect Direction of Rotation	●		●			●	●			
Insufficient Belt Tension							●	●		
Loose Dampers or Variable-inlet Vanes							●			
Misalignment of Bearings, Coupling, Wheel, or Belts				●		●	●	●	●	
Motor Improperly Wired						●	●	●		●
Packing Too Tight or Defective Stuffing Box						●	●		●	●
Poor Fan Inlet or Outlet Conditions	●		●							
Specific Gravity or Density Above Design						●	●		●	
Speed Too High		●		●	●	●	●			●
Speed Too Low	●	●	●					●		●
Too Much Grease in Ball Bearings				●						
Total System Head Greater Than Design	●		●	●		●			●	
Total System Head Less Than Design		●					●			●
Unstable Foundation		●		●			●	●		
Vibration Transmitted to Fan from Outside Sources				●			●	●		
Wheel Binding on Fan Housing				●		●	●	●		●
Wheel Mounted Backward on Shaft	●		●							
Worn Bearings							●	●		
Worn Coupling							●			
120-Cycle Magnetic Hum							●	●		

Table 15.4 Common Failure Modes of Blowers and Fluidizers

THE CAUSES	No Air/Gas Delivery	Insufficient Discharge Pressure	Insufficient Capacity	Excessive Wear	Excessive Heat	Excessive Vibration and Noise	Excessive Power Demand	Motor Trips	Elevated Motor Temperature	Elevated Air/Gas Temperature
Air Leakage into Suction Piping or Shaft Seal		●	●			●				
Coupling Misaligned				●	●	●	●		●	
Excessive Discharge Pressure			●	●		●	●	●		●
Excessive Inlet Temperature/Moisture			●							
Insufficient Suction Air/Gas Supply		●	●	●		●			●	
Internal Component Wear	●	●	●							
Motor or Driver Failure	●									
Pipe Strain on Blower Casing				●	●	●	●		●	
Relief Valve Stuck Open or Set Wrong		●	●							
Rotating Element Binding				●	●	●	●	●	●	
Solids or Dirt in Inlet Air/Gas Supply				●						
Speed Too Low		●	●						●	
Suction Filter or Strainer Clogged	●	●	●			●			●	
Wrong Direction of Rotation	●	●							●	

assembly and shaft to deflect from their true centerline. This deflection increases the vibration energy of the fan and accelerates the wear rate of bearings and other drive-train components.

Plate-Out

Dirt, moisture, and other contaminants tend to adhere to the fan's rotating element. This buildup, called *plate-out*, increases the mass of the rotor assembly and decreases its critical speed, the point at which the phenomenon referred to as *resonance* occurs. This occurs because the additional mass affects the rotor's natural frequency. Even if the fan's speed does not change, the change in natural frequency may cause its critical speed (note that machines may have more than one) to coincide with the actual rotor speed. If this occurs, the fan will resonate, or experience severe vibration, and may catastrophically fail. The symptoms of plate-out are often confused with those of mechanical imbalance because both dramatically increase the vibration associated with the fan's running speed.

The problem of plate-out can be resolved by regularly cleaning the fan's rotating element and internal components. Removal of buildup lowers the rotor's mass and returns its natural frequency to the initial or design point. In extremely dirty or dusty environments, it may be advisable to install an automatic cleaning system that uses high-pressure air or water to periodically remove any buildup that occurs.

Speed Changes

In applications in which a measurable fan speed change can occur (i.e., V-belt or variable-speed drives), care must be taken to ensure that the selected speed does not coincide with any of the fan's critical speeds. For general-purpose fans, the actual running speed is designed to be between 10% and 15% below the first critical speed of the rotating element. If the sheave ratio of a V-belt drive or the actual running speed is increased above the design value, it may coincide with a critical speed.

Some fans are designed to operate between critical speeds. In these applications, the fan must transition through the first critical speed to reach its operating speed. These transitions must be made as quickly as possible to prevent damage. If the fan's speed remains at or near the critical speed for any extended period of time, serious damage can occur.

Lateral Flexibility

By design, the structural support of most general-purpose fans lacks the mass and rigidity needed to prevent flexing of the fan's housing and rotating assembly. This problem is more pronounced in the horizontal plane, but it also is present in the vertical direction. If support-structure flexing is found to be the root cause or a major contributing factor to the problem, it can be corrected by increasing the stiffness and/or mass of the structure. However, do not fill the structure with concrete. As it dries, concrete pulls away from the structure and does little to improve its rigidity.

BLOWERS OR POSITIVE-DISPLACEMENT FANS

Blowers, or positive-displacement fans, have the same common failure modes as rotary pumps and compressors. Table 15.4 lists the failure modes that most often affect blowers and fluidizers. In particular, blower failures occur because of process instability caused by start/stop operation and demand variations and because of mechanical failures caused by close tolerances.

Process Instability

Blowers are very sensitive to variations in their operating envelope. As little as a 1-psig change in downstream pressure can cause the blower to become extremely unstable. The probability of catastrophic failure or severe damage to blower components increases in direct proportion to the amount and speed of the variation in demand or downstream pressure.

Start/Stop Operation

The transients caused by frequent start/stop operation also have a negative effect on blower reliability. Conversely, blowers that operate constantly in a stable environment rarely exhibit problems. The major reason is the severe axial thrusting caused by the frequent variations in suction or discharge pressure caused by the start/stop operation.

Demand Variations

Variations in pressure and volume demands have a serious impact on blower reliability. Since blowers are positive-displacement devices, they generate a constant volume and a variable pressure that is dependent on the downstream system's backpressure. If demand decreases, the blower's discharge pressure continues to increase until (1) a downstream component fails and reduces the backpressure or (2) the brake horsepower required to drive the blower is greater than the motor's locked rotor rating. Either of these results in failure of the blower system. The former may result in a reportable release, while the latter will cause the motor to trip or burnout.

Frequent variations in demand greatly accelerate the wear rate of the thrust bearings in the blower. This can be directly attributed to the constant, instantaneous axial thrusting caused by variations in the discharge pressure required by the downstream system.

Mechanical Failures

Because of the extremely close clearances that must exist within the blower, the potential for serious mechanical damage or catastrophic failure is higher than with other rotating machinery. The primary failure points include thrust bearing, timing gears, and rotor assemblies.

In many cases, these mechanical failures are caused by the instability discussed in the preceding sections, but poor maintenance practices are another major cause.

16

DUST COLLECTORS

The basic operations performed by dust-collection devices are (1) separating particles from the gas stream by deposition on a collection surface; (2) retaining the deposited particles on the surface until removal; and (3) removing the deposit from the surface for recovery or disposal.

The separation step requires the following: (1) application of a force that produces a differential motion of the particles relative to the gas, and (2) sufficient gas-retention time for the particles to migrate to the collecting surface. Most dust-collection systems are composed of a pneumatic conveying system and some device that separates suspended particulate matter from the conveyed air stream. The more common systems use either filter media (e.g., fabric bags) or cyclonic separators to separate the particulate matter from air.

BAGHOUSES

Fabric-filter systems, commonly called *bag-filter* or *baghouse* systems, are dust-collection systems in which dust-laden air is passed through a bag-type filter. The bag collects the dust in layers on its surface and the dust layer itself effectively becomes the filter medium. Because the bag's pores are usually much larger than those of the dust-particle layer that forms, the initial efficiency is very low. However, it improves once adequate dust-layer forms. Therefore, the potential for dust penetration of the filter media is extremely low except during the initial period after startup, bag change, or during the fabric-cleaning, or blow-down, cycle.

The principal mechanisms of disposition in dust collectors are (1) gravitational deposition, (2) flow-line interception, (3) inertial deposition, (4) diffusion deposition, and (5) electrostatic deposition. During the initial operating period, particle deposition takes place mainly by inertial and flow-line interception, diffusion, and gravity. Once the dust layer has been fully established, sieving is probably the dominant deposition mechanism.

Configuration

A baghouse system consists of the following: pneumatic-conveyor system, filter media, a back-flush cleaning system, and a fan or blower to provide airflow.

Pneumatic Conveyor

The primary mechanism for conveying dust-laden air to a central collection point is a system of pipes or ductwork that functions as a pneumatic conveyor. This system gathers dust-laden air from various sources within the plant and conveys it to the dust-collection system.

Dust-Collection System

The design and configuration of the dust-collection system varies with the vendor and the specific application. Generally, a system consists of either a single large hopper-like vessel or a series of hoppers with a fan or blower affixed to the discharge manifold. Inside the vessel is an inlet manifold that directs the incoming air or gas to the dirty side of the filter media or bag. A plenum, or divider plate, separates the dirty and clean sides of the vessel.

Filter media, usually long cylindrical tubes or bags, are attached to the plenum. Depending on the design, the dust-laden air or gas may flow into the cylindrical filter bag and exit to the clean side or it may flow through the bag from its outside and exit through the tube's opening. Figure 16.1 illustrates a typical baghouse configuration.

Fabric-filter designs fall into three types, depending on the method of cleaning used: (1) shaker-cleaned, (2) reverse-flow-cleaned, and (3) reverse-pulse-cleaned.

Shaker-Cleaned Filter The open lower ends of shaker-cleaned filter bags are fastened over openings in the tube sheet that separates the lower, dirty-gas inlet chamber from the upper clean-gas chamber. The bags are suspended from supports, which are connected to a shaking device.

The dirty gas flows upward into the filter bag and the dust collects on the inside surface. When the pressure drop rises to a predetermined upper limit because of dust accumulation, the gas flow is stopped and the shaker is operated. This process dislodges the dust, which falls into a hopper located below the tube sheet.

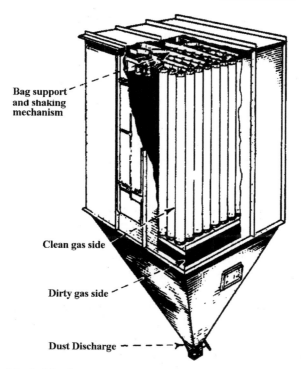

Bag support
and shaking
mechanism

Clean gas side

Dirty gas side

Dust Discharge

Figure 16.1 A typical baghouse.

For continuous operation, the filter must be constructed with multiple compartments. This is necessary so that individual compartments can be sequentially taken off line for cleaning while the other compartments continue to operate.

Ordinary shaker-cleaned filters may be cleaned every 15 minutes to 8 hours, depending on the service conditions. A manometer connected across the filter is used to determine the pressure drop, which indicates when the filter should be shaken. Fully automatic filters may be shaken every 2 minutes, but bag maintenance is greatly reduced if the time between shakings can be increased to 15 to 20 minutes.

The determining factor in the frequency of cleaning is the pressure drop. A differential-pressure switch can serve as the actuator in automatic cleaning applications. Cyclone pre-cleaners are sometimes used to reduce the dust load on the filter or to remove large particles before they enter the bag.

It is essential to stop the gas flow through the filter during shaking for the dust to fall off. With very fine dust, it may be necessary to equalize the pressure across the cloth. In practice, this can be accomplished without interrupting continuous operation by removing one section from service at a time. With automatic filters, this operation involves closing the dirty-gas inlet dampers, shaking the filter units either pneumatically or mechanically, and reopening the dampers. In some cases, a reverse flow of clean gas through the filter is used to augment the shaker-cleaning process.

The gas entering the filter must be kept above its dew point to avoid water-vapor condensation on the bags, which will cause plugging. However, fabric filters have been used successfully in steam atmospheres, such as those encountered in vacuum dryers. In these applications, the housing is generally steam-cased.

Reverse-Flow-Cleaned Filter Reverse-flow-cleaned filters are similar to the shaker-cleaned design, except the shaker mechanism is eliminated. As with shaker-cleaned filters, compartments are taken off line sequentially for cleaning. The primary use of reverse-flow cleaning is in units that use fiberglass-fabric bags at temperatures above 150°C (300°F).

After the dirty-gas flow is stopped, a fan forces clean gas through the bags from the clean-gas side. The superficial velocity of the gas through the bag is generally 1.5–2.0 feet per minute, or about the same velocity as the dirty-gas inlet flow. This flow of clean gas partially collapses the bag and dislodges the collected dust, which falls to the hopper. Rings are usually sewn into the bags at intervals along their length to prevent complete collapse, which would obstruct the fall of the dislodged dust.

Reverse-Pulse-Cleaned Filter In the reverse-pulse-cleaned filter, the bag forms a sleeve drawn over a cylindrical wire cage, which supports the fabric on the clean-gas side (i.e., inside) of the bag. The dust collects on the outside of the bag.

A venturi nozzle is located in the clean-gas outlet from each bag, which is used for cleaning. A jet of high-velocity air is directed through the venturi nozzle and into the bag, which induces clean gas to pass through the fabric to the dirty side. The high-velocity jet is released in a short pulse, usually about 100 milliseconds, from a compressed air line by a solenoid-controlled valve. The pulse of air and clean gas expand the bag and dislodge the collected dust. Rows of bags are cleaned in a timed sequence by programmed operation of the solenoid valves. The pressure of the pulse must be sufficient to dislodge the dust without cessation of gas flow through the baghouse.

It is common practice to clean the bags on-line without stopping the flow of dirty gas into the filter. Therefore, reverse-pulse bag filters are often built without multiple compartments. However, investigations have shown that a large fraction of the dislodged dust redeposits on neighboring bags rather than falling to the dust hopper.

As a result, there is a growing trend to off-line clean reverse-pulse filters by using bags with multiple compartments. These sections allow the outlet-gas plenum serving a particular section to be closed off from the clean-gas exhaust, thereby stopping the flow of inlet gas. On the dirty side of the tube sheet, the isolated section is separated by partitions from the neighboring sections in which filtration continues. Sections of the filter are cleaned in rotation as with shaker and reverse-flow filters.

Some manufacturers design bags for use with relatively low-pressure air (i.e., 15 psi) instead of the normal 100 psi air. This allows them to eliminate the venturi tubes for clean-gas induction. Others have eliminated the separate jet nozzles located at the individual bags in favor of a single jet to pulse air into the outlet-gas plenum.

Reverse-pulse filters are typically operated at higher filtration velocities (i.e., air-to-cloth ratios) than shaker or reverse-flow designs. Filtration velocities may range from 3–15 feet per minute in reverse-pulse applications, depending on the dust being collected. However, the most the commonly used range is 4–5 feet per minute.

The frequency of cleaning depends on the nature and concentration of the dust. Typical cleaning intervals vary from about 2 to 15 minutes. However, the cleaning action of the pulse is so effective that the dust layer may be completely removed from the surface of the fabric. Consequently, the fabric itself must serve as the principal filter medium for a substantial part of the filtration cycle, which decreases cleaning efficiency. Because of this, woven fabrics are unsuitable for use in these devices and felt-type fabrics are used instead. With felt filters, although the bulk of the dust is still removed, the fabric provides an adequate level of dust collection until the dust layer reforms.

Cleaning System

As discussed in the preceding section, filter bags must be periodically cleaned to prevent excessive buildup of dust and to maintain an acceptable pressure drop across the filters. Two of the three designs discussed, reverse-flow and reverse-pulse, depend on an adequate supply of clean air or gas to provide this periodic cleaning. Two factors are critical in these systems: the clean-gas supply and the proper cleaning frequency.

Clean-Gas Supply Most applications that use the reverse-flow cleaning system use ambient air as the primary supply of clean gas. A large fan or blower draws ambient air into the clean side of the filter bags. However, unless inlet filters properly condition the air, it may contain excessive dirt loads that can affect the bag life and efficiency of the dust-collection system.

In reverse-pulse applications, most plants rely on plant-air systems as the source for the high-velocity pulses required for cleaning. In many cases, however, the plant-air system is not sufficient for this purpose. Although the pulses required are short (i.e., 100 milliseconds or less), the number and frequency can deplete the supply. Therefore, care must be taken to ensure that both sufficient volume and pressure are available to achieve proper cleaning.

Cleaning Frequency Proper operation of a baghouse, regardless of design, depends on frequent cleaning of the filter media. The system is designed to operate within a specific range of pressure drops that defines clean and fully loaded filter media. The cleaning frequency must ensure that the maximum recommended pressure drop is not exceeded.

This can be a real problem for baghouses that rely on automatic timers to control cleaning frequency. The use of a timing function to control cleaning frequency is not recommended unless the dust load is known to be consistent. A better approach is to use differential-pressure gauges to physically measure the pressure drop across the filter media to trigger the cleaning process based on preset limits.

Fan or Blower

All baghouse designs use some form of fan, blower, or centrifugal compressor to provide the dirty-air flow required for proper operation. In most cases, these units are installed on the clean side of the baghouse to draw the dirty air through the filter media.

Since these units provide the motive power required to transport and collect the dust-laden air, their operating condition is critical to the baghouse system. The type and size of air-moving unit varies with the baghouse type and design. Refer to the O&M manuals, as well as Chapter 2 (Fans and Blowers) or Chapter 4 (Compressors) for specific design criteria for these critical units.

Performance

The primary measure of baghouse-system performance is its ability to consistently remove dust and other particulate matter from the dirty-air stream. Pressure drop and collection efficiency determine the effectiveness of these systems.

Pressure Drop

The filtration, or superficial face velocities used in fabric filters are generally in the range of 1–10 feet per minute, depending on the type of fabric, fabric supports, and cleaning methods used. In this range, pressure drops conform to Darcy's law for streamline flow in porous media, which states that the pressure drop is directly proportional to the flow rate. The pressure drop across the fabric media and the dust layer may be expressed by:

$$\Delta p = K_1 V_f + K_2 \omega V_f$$

where

Δp = Pressure drop (inches of water)

Vf = Superficial velocity through filter (ft/min)

ω = Dust loading on filter (lbm/ft^2)

K_1 = Resistance coefficient for conditioned fabric
(inches of water/ft/min)

K_2 = Resistance coefficient for dust layer
(inches of water/lbm/ft/min)

Conditioned fabric maintains a relatively consistent dust-load deposit following a number of filtration and cleaning cycles. K_1 may be more than 10 times the value of the resistance coefficient for the original clean fabric. If the depth of the dust layer on the fabric is greater than about 1/16 in. (which corresponds to a fabric dust loading on the order of 0.1 lbm/ft^2), the pressure drop across the fabric, including the dust in the pores, is usually negligible relative to that across the dust layer alone.

In practice, K_1 and K_2 are measured directly in filtration experiments. These values can be corrected for temperature by multiplying by the ratio of the gas viscosity at the desired condition to the gas viscosity at the original experimental condition.

Collection Efficiency

Under controlled conditions (e.g., in the laboratory), the inherent collection efficiency of fabric filters approaches 100%. In actual operation, it is determined by several variables, in particular the properties of the dust to be removed, choice of filter fabric, gas velocity, method of cleaning, and cleaning cycle. Inefficiency usually results from bags that are poorly installed, torn, or stretched from excessive dust loading and excessive pressure drop.

Installation

Most baghouse systems are provided as complete assemblies by the vendor. While the unit may require some field assembly, the vendor generally provides the structural supports, which in most cases are adequate. The only controllable installation factors that may affect performance are the foundation and connections to pneumatic conveyors and other supply systems.

Foundation

The foundation must support the weight of the baghouse. In addition, it must absorb the vibrations generated by the cleaning system. This is especially true when using the shaker-cleaning method, which can generate vibrations that can adversely affect the structural supports, foundation, and adjacent plant systems.

Connections

Efficiency and effectiveness depend on leak-free connections throughout the system. Leaks reduce the system's ability to convey dust-laden air to the baghouse. One potential source for leaks is improperly installed filter bags. Because installation varies with the type of bag and baghouse design, consult the vendor's O&M manual for specific instructions.

Operating Methods

The guidelines provided in the vendor's O&M manual should be the primary reference for proper baghouse operation. Vendor-provided information should be used because there are not many common operating guidelines among the various configurations. The only general guidelines that are applicable to most designs are cleaning frequency and inspection and replacement of filter media.

Cleaning

As previously indicated, most bag-type filters require a pre-coat of particulates before they can effectively remove airborne contaminants. However, particles can completely block air flow if the filter material becomes overloaded. Therefore the primary operating criterion is to maintain the efficiency of the filter media by controlling the cleaning frequency.

Most systems use a time sequence to control the cleaning frequency. If the particulate load entering the baghouse is constant, this approach would be valid. However, the incoming load generally changes constantly. As a result, the straight time sequence methodology does not provide the most efficient mode of operation.

Operators should monitor the differential-pressure gauges that measure the total pressure drop across the filter media. When the differential pressure reaches the maximum recommended level (data provided by the vendor), the operator should override any automatic timer controls and initiate the cleaning sequence.

Inspecting and Replacing Filter Media

Filter media used in dust-collection systems are prone to damage and abrasive wear. Therefore, regular inspection and replacement is needed to ensure continuous, long-term performance. Any damaged, torn, or improperly sealed bags should be removed and replaced.

One of the more common problems associated with baghouses is improper installation of filter media. Therefore it is important to follow the instructions provided by the vendor. If the filter bags are not properly installed and sealed, overall efficiency and effectiveness are significantly reduced.

CYCLONE SEPARATORS

A widely used type of dust-collection equipment is the cyclone separator. A cyclone is essentially a settling chamber in which gravitational acceleration is replaced by centrifugal acceleration. Dust-laden air or gas enters a cylindrical or conical chamber tangentially at one or more points and leaves through a central opening. The dust particles, by virtue of their inertia, tend to move toward the outside separator wall from where they are led into a receiver. Under common operating conditions, the centrifugal separating force or acceleration may range from five times gravity in very large diameter, low-resistance cyclones to 2,500 times gravity in very small, high-resistance units.

Within the range of their performance capabilities, cyclones are one of the least expensive dust-collection systems. Their major limitation is that, unless very small units are used, efficiency is low for particles smaller than 5 microns. Although cyclones may be used to collect particles larger than 200 microns, gravity-settling chambers or simple inertial separators are usually satisfactory and less subject to abrasion.

Configuration

The internal configuration of a cyclone separator is relatively simple. Figure 16.2 illustrates a typical cross-section of a cyclone separator, which consists of the following segments:

- Inlet area that causes the gas to flow tangentially
- Cylindrical transition area

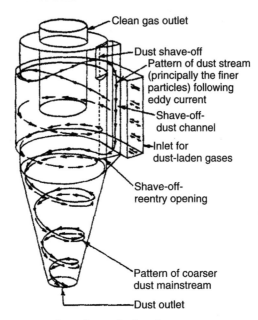

Figure 16.2 Flow pattern through a typical cyclone separator.

- Decreasing taper that increases the air velocity as the diameter decreases
- Central return tube to direct the dust-free air out the discharge port.

Particulate material is forced to the outside of the tapered segment and collected in a drop-leg located at the dust outlet. Most cyclones have a rotor-lock valve affixed to the bottom of the drop-leg. This is a motor-driven valve that collects the particulate material and discharges it into a disposal container.

Performance

Performance of a cyclone separator is determined by flow pattern, pressure drop, and collection efficiency.

Flow Pattern

The path the gas takes in a cyclone is through a double vortex that spirals the gas downward at the outside and upward at the inside. When the gas enters the cyclone, the tangential component of its velocity, V_{ct}, increases with the decreasing radius as expressed by:

$$V_{ct} \approx r^{-n}$$

In this equation, r is the cyclone radius and n is dependent on the coefficient of friction. Theoretically, in the absence of wall friction, n should equal 1.0. Actual measurements, however, indicate that n ranges from 0.5 to 0.7 over a large portion of the cyclone radius. The spiral velocity in a cyclone may reach a value several times the average inlet gas velocity.

Pressure Drop

The pressure drop and the friction loss through a cyclone are most conveniently expressed in terms of the velocity head based on the immediate inlet area. The inlet velocity head, h_{vt}, which is expressed in inches of water, is related to the average inlet gas velocity and density by:

$$h_{vt} = 0.0030\rho V_c^2$$

where

h_{vt} = Inlet-velocity head (inches of water)
ρ = Gas density (lb/ft^3)
V_c = Average inlet gas velocity (ft/sec)

The cyclone friction loss, F_{cv}, is a direct measure of the static pressure and power that a fan must develop. It is related to the pressure drop by:

$$F_{cv} = \Delta p_{cv} + 1 - \left(\frac{4A_c}{\pi D_e^2}\right)^2$$

where:

F_{cv} = Friction loss (inlet-velocity heads)
Δp_{cv} = Pressure drop through the cyclone (inlet-velocity heads)
A_c = Area of the cyclone (ft.^{2})
D_e = Diameter of the gas exit (ft.)

The friction loss through cyclones may range from 1 to 20 inlet-velocity heads, depending on its geometric proportions. For a cyclone of specific geometric proportions, F_{cv} and Δp_{cv}, are essentially constant and independent of the actual cyclone size.

Collection Efficiency

Since cyclones rely on centrifugal force to separate particulates from the air or gas stream, particle mass is the dominant factor that controls efficiency. For particulates with high densities (e.g., ferrous oxides), cyclones can achieve 99% or better removal efficiencies, regardless of particle size. Lighter particles (e.g., tow or flake) dramatically reduce cyclone efficiency.

These devices are generally designed to meet specific pressure-drop limitations. For ordinary installations operating at approximately atmospheric pressure, fan limitations dictate a maximum allowable pressure drop corresponding to a cyclone inlet velocity in the range of 20–70 feet per second. Consequently, cyclones are usually designed for an inlet velocity of 50 feet per second.

Varying operating conditions change dust-collection efficiency only by a small amount. The primary design factor that controls collection efficiency is cyclone diameter. A small-diameter unit operating at a fixed pressure drop has a higher efficiency than a large-diameter unit. Reducing the gas-outlet duct diameter also increases the collection efficiency.

Installation

As in any other pneumatic-conveyor system, special attention must be given to the piping or ductwork used to convey the dust-laden air or gas. The inside surfaces must be smooth and free of protrusions that affect the flow pattern. All bends should be gradual and provide a laminar-flow path for the gas.

TROUBLESHOOTING

This section identifies common problems and their causes for baghouse and cyclonic separator dust-collection systems.

BAGHOUSES

Table 16.1 lists the common failure modes for baghouses. This guide may be used for all such units that use fabric filter bags as the primary dust-collection media.

CYCLONIC SEPARATORS

Table 16.2 identifies the failure modes and their causes for cyclonic separators. Since there are no moving parts within a cyclone, most of the problems associated with this type of system can be attributed to variations in process parameters such as flow rate, dust load, dust composition (i.e., density, size, etc.), and ambient conditions (i.e., temperature, humidity, etc.).

Table 16.1 Common Failure Modes of Baghouses

THE CAUSES	Continuous Release of Dust-laden Air	Intermittent Release of Dust-laden Air	Loss of Plant Air Pressure	Blow-down Ineffective	Insufficient Capacity	Excessive Differential Pressure	Fan/Blower Motor Trips	Fan Has High Vibration	Premature Bag Failures	Differential Pressure Too Low	Chronic Plugging of Bags
Bag Material Incompatible For Application									●		●
Bag Plugged						●	●	●			
Bag Torn or Improperly Installed	●							●	●	●	
Baghouse Undersized					●			●			●
Blow-down Cycle Interval Too Long						●	●				
Blow-down Cycle Time Failed or Damaged						●	●				
Blow-down Nozzles Plugged						●					
Blow-down Pilot Valve Failed To Open (Solenoid Failure)		●				●					
Dust Load Exceeds Capacity											●
Excessive Demand			●								
Fan/Blower Not Operating Properly					●						
Improper or Inadequate Lubrication								●			
Leaks In Ductwork or Baghouse	●				●						
Misalignment of Fan and Motor								●			
Moisture Content Too High											●
Not Enough Blow-down Air (Pressure and Volume)			●	●		●					
Not Enough Dust Layer on Filter Bags	●	●							●	●	
Piping/Valve Leaks			●								
Plate-out (Dust Buildup on Fan's Rotor)								●			
Plenum Cracked or Seal Defective	●		●							●	
Rotor Imbalanced								●			
Ruptured Blow-down Diaphrag			●	●		●					
Suction Ductwork Blocked or Plugged					●						

Table 16.2 Common Failure Modes of Cyclonic Separators

THE CAUSES	Continuous Release of Dust-laden Air	Intermittent Release of Dust-laden Air	Cyclone Plugs in Inlet Chamber	Cyclone Plugs in Dust Removal Section	Rotor-lock Valve Fails To Turn	Excessive Differential Pressure	Differential Pressure Too Low	Rotor-lock Valve Leaks	Fan Has High Vibration
Clearance Set Wrong								●	
Density and Size Distribution of Dust Too High				●	●	●			●
Density and Size Distribution of Dust Too Low	●	●							
Dust Load Exceeds Capacity	●	●			●				●
Excessive Moisture in Incoming Air			●						
Foreign Object Lodged in Valve					●				
Improper Drive-train Adjustments					●				
Improper Lubrication					●				
Incoming Air Velocity Too High						●			
Incoming Air Velocity Too Low	●	●	●				●		
Internal Wear or Damage								●	
Large Contaminates in Incoming Air Stream			●		●				
Prime Mover (Fan, Blower) Malfunctioning	●	●				●	●		●
Rotor-lock Valve Turning Too Slow	●	●		●					
Seals Damaged								●	

17

PUMPS

CENTRIFUGAL PUMPS

Centrifugal pumps basically consist of a stationary pump casing and an impeller mounted on a rotating shaft. The pump casing provides a pressure boundary for the pump and contains channels to properly direct the suction and discharge flow. The pump casing has suction and discharge penetrations for the main flow path of the pump and normally has a small drain and vent fittings to remove gases trapped in the pump casing or to drain the pump casing for maintenance.

Figure 17.1 is a simplified diagram of a typical centrifugal pump that shows the relative locations of the pump suction, impeller, volute, and discharge. The pump casing guides the liquid from the suction connection to the center, or eye, of the impeller. The vanes of the rotating impeller impart a radial and rotary motion to the liquid, forcing it to the outer periphery of the pump casing, where it is collected in the outer part of the pump casing, called the *volute*.

The volute is a region that expands in cross-sectional area as it wraps around the pump casing. The purpose of the volute is to collect the liquid discharged from the periphery of the impeller at high velocity and gradually cause a reduction in fluid velocity by increasing the flow area. This converts the velocity head to static pressure. The fluid is then discharged from the pump through the discharge connection. Figure 17.2 illustrates the two types of volutes.

Centrifugal pumps can also be constructed in a manner that results in two distinct volutes, each receiving the liquid that is discharged from a 180-degree region of the impeller at any given time. Pumps of this type are called *double*

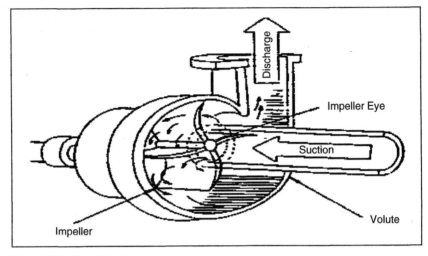

Figure 17.1 Centrifugal pump.

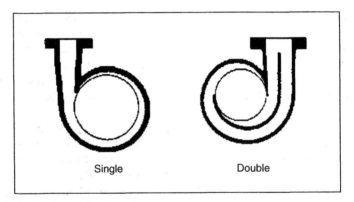

Figure 17.2 Single and double volute.

volute pumps. In some applications the double volute minimizes radial forces imparted to the shaft and bearings because of imbalances in the pressure around the impeller.

CHARACTERISTICS CURVE

For a given centrifugal pump operating at a constant speed, the flow rate through the pump is dependent on the differential pressure or head developed by the pump. The lower the pump head, the higher the flow rate. A vendor manual for a specific pump usually contains a curve of pump flow rate versus pump head called a *pump characteristic* curve. After a pump is installed in a system, it is usually tested to ensure that the flow rate and head of the pump are within the required specifications. A typical centrifugal pump characteristic curve is shown in Figure 17.3.

Several terms associated with the pump characteristic curve must be defined. *Shutoff head* is the maximum head that can be developed by a centrifugal pump operating at a set speed. *Pump run-out* is a the maximum flow a centrifugal pump can develop without damaging the pump. Centrifugal pumps must be designed to be protected from the conditions of pump run-out or operating at shutoff head.

PROTECTION

A centrifugal pump is deadheaded when it is operated with a closed discharge valve or against a seated check valve. If the discharge valve is closed and there is no other flow path available to the pump, the impeller will churn the same volume of water as it rotates in the pump casing. This will increase the temperature of the liquid in the pump casing to the point that it will flash to vapor. If the pump is run in this condition for a significant amount of time, it will become damaged.

When a centrifugal pump is installed in a system such that it may be subjected to periodic shutoff head conditions, it is necessary to provide some means of pump protection. One method for protecting the pump from running deadheaded is to

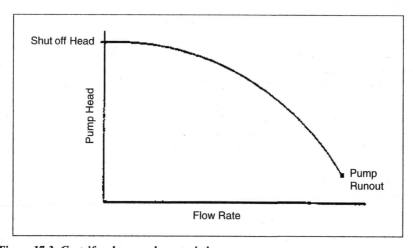

Figure 17.3 Centrifugal pump characteristic curve.

provide a recirculation line from the pump discharge line upstream of the discharge valve back to the pump's supply source. The recirculation line should be sized to allow enough flow through the pump to prevent overheating and damage to the pump. Protection may also be accomplished by use of an automatic flow control device.

Centrifugal pumps must also be protected from run-out. One method for ensuring that there is always adequate flow resistance at the pump discharge to prevent excessive flow through the pump is to place an orifice or a throttle valve immediately downstream of the pump discharge.

Gas Binding

Gas binding of a centrifugal pump is a condition in which the pump casing is filled with gases or vapors to the point where the impeller is no longer able to contact enough fluid to function correctly. The impeller spins in the gas bubble but is unable to force liquid through the pump.

Centrifugal pumps are designed so that their pump casings are completely filled with liquid during pump operation. Most centrifugal pumps can still operate when a small amount of gas accumulates in the pump casing, but pumps in systems containing dissolved gases that are not designed to be self-venting should be periodically vented manually to ensure that gases do not build up in the pump casing.

Priming

Most centrifugal pumps are not self-priming. In other words, the pump casing must be filled with liquid before the pump is started or the pump will not be able to function. If the pump casing becomes filled with vapors or gases, the pump impeller becomes gas-bound and incapable of pumping. To ensure that a centrifugal pump remains primed and does not become gas-bound, most centrifugal pumps are located below the level of the source from which the pump is to take its suction. The same effect can be gained by supplying liquid to the pump suction under pressure supplied by another pump placed in the suction line.

Classification by Flow

Centrifugal pumps can be classified based on the manner in which fluid flows through the pump. The manner in which fluid flows through the pump is determined by the design of the pump casing and the impeller. The three types of flow through a centrifugal pump are radial flow, axial flow, and mixed flow.

Radial Flow

In a radial flow pump, the liquid enters at the center of the impeller and is directed out along the impeller blades in a direction at right angles to the pump shaft. The impeller of a typical radial flow pump and the flow is illustrated in Figure 17.4.

Axial Flow

In an axial flow pump, the impeller pushes the liquid in a direction parallel to the pump shaft. Axial flow pumps are sometimes called *propeller* pumps because they operate essentially the same as the propeller of a boat. The impeller of a typical axial flow pump and the flow through a radial flow pump are shown in Figure 17.5.

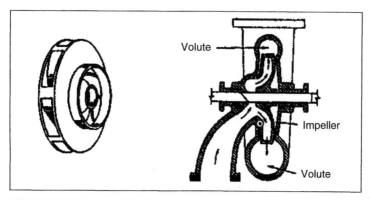

Figure 17.4 Radial flow centrifugal pump.

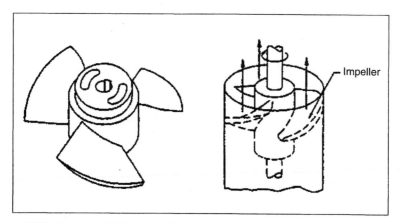

Figure 17.5 Typical axial flow centrifugal pump.

Mixed Flow

Mixed flow pumps borrow characteristics from both radial flow and axial flow pumps. As liquid flows through the impeller of a mixed flow pump, the impeller blades push the liquid out away from the pump shaft and to the pump suction at an angle greater than 90 degrees. The impeller of a typical mixed flow pump and the flow through a mixed flow pump are shown in Figure 17.6.

MULTI-STAGE PUMPS

A centrifugal pump with a single impeller that can develop a differential pressure of more than 150 psid between the suction and the discharge is difficult and costly to design and construct. A more economical approach to developing high pressures with a single centrifugal pump is to include multiple impellers on a common shaft within the same pump casing. Internal channels in the pump casing route the discharge of one impeller to the suction of another impeller. Figure 17.7 shows a diagram of the arrangement of the impellers of a four-stage pump. The water enters the pump from the top left and passes through each of the four impellers, going from left to right. The water goes from the volute surrounding the discharge of one impeller to the suction of the next impeller.

A pump stage is defined as that portion of a centrifugal pump consisting of one impeller and its associated components. Most centrifugal pumps are single-stage pumps, containing only one impeller. A pump containing seven impellers within a single casing would be referred to as a *seven-stage* pump or generally as a multi-stage pump.

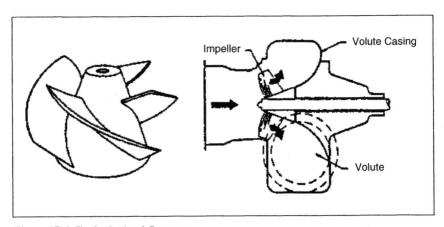

Figure 17.6 Typical mixed flow pump.

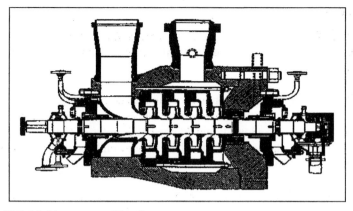

Figure 17.7 Multi-stage centrifugal pump.

COMPONENTS

Centrifugal pumps vary in design and construction from simple pumps with relatively few parts to extremely complicated pumps with hundreds of individual parts. Some of the most common components found in centrifugal pumps are wearing rings, stuffing boxes, packing, and lantern rings. These components are shown in Figure 17.8 and are described in the following pages.

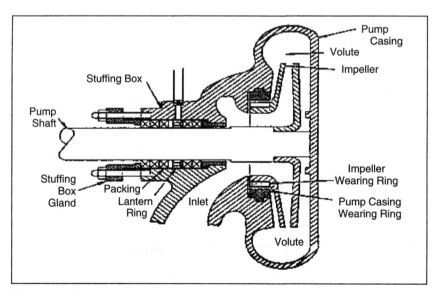

Figure 17.8 Components of a centrifugal pump.

Impellers

Impellers of pumps are classified based on the number of points that the liquid can enter the impeller and also on the amount of webbing between the impeller blades.

Impellers can be either single-suction or double-suction. A single-suction impeller allows liquid to enter the center of the blades from only one direction. A double-suction impeller allows liquid to enter the center of the impeller blades from both sides simultaneously. Figure 17.9 shows simplified diagrams of single- and double-suction impellers.

Impellers can be open, semi-open, or enclosed. The open impeller consists only of blades attached to a hub. The semi-open impeller is constructed with a circular plate (the web) attached to one side of the blade. The enclosed impeller has circular plates attached to both sides of the blades. Enclosed impellers are also referred to as *shrouded* impellers. Figure 17.10 illustrates examples of open, semi-open, and enclosed impellers.

The impeller sometimes contains balancing holes that connect the space around the hub to the suction side of the impeller. The balancing holes have a total cross-sectional area that is considerably greater than the cross-sectional area of the annular space between the wearing ring and the hub. The result is suction pressure on both sides of the impeller hub, which maintains a hydraulic balance of axial thrust.

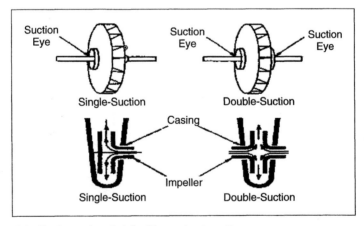

Figure 17.9 Single-suction and double-suction impellers.

Figure 17.10 Open, semi-open, and enclosed impellers.

Diffuser

Some centrifugal pumps contain diffusers. A diffuser is a set of stationary vanes that surround the impeller. The purpose of the diffuser is to increase the efficiency of the centrifugal pump by allowing a more gradual expansion and less turbulent area for the liquid to reduce in velocity. The diffuser vanes are designed in a manner that the liquid exiting the impeller will encounter an ever-increasing flow area as it passes through the diffuser. This increase in flow area causes a reduction in flow velocity, converting kinetic energy into flow energy. The increase in flow energy can be observed as an increase in the pressure of an incompressible fluid. Figure 17.11 shows a centrifugal pump diffuser.

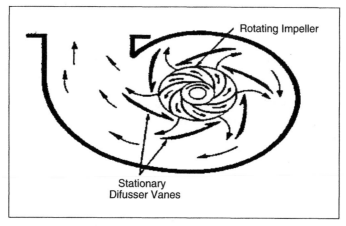

Figure 17.11 Centrifugal pump diffuser.

Wearing Rings

Centrifugal pumps contain rotating impellers within stationary pump casings. To allow the impeller to rotate freely within the pump casing, a small clearance is maintained between the impeller and the pump casing. To maximize the efficiency of a centrifugal pump, it is necessary to minimize the amount of liquid leaking through this clearance from the high-pressure side or discharge side of the pump back to the low-pressure or suction side.

It is unavoidable that some wear will occur at the point where the impeller and the pump casing nearly come into contact. This wear is due to the erosion caused by liquid leaking through this tight clearance and other causes. Eventually the leakage could become unacceptably large and maintenance would be required on the pump.

To minimize the cost of pump maintenance, many centrifugal pumps are designed with wearing rings. Wearing rings are replaceable rings that are attached to the impeller and/or the pump casing to allow a small running clearance between the impeller and pump casing without causing wear of the actual impeller or pump casing material.

Stuffing Box

In almost all centrifugal pumps, the rotating shaft that drives the impeller penetrates the pressure boundary of the pump casing. It is important that the pump is designed properly to control the amount of liquid that leaks along the shaft at the point that the shaft penetrates the pump casing. Factors considered when choosing a method include the pressure and temperature of the fluid being pumped, the size of the pump, and the chemical and physical characteristics of the fluid being pumped.

One of the simplest types of shaft seal is the stuffing box. The stuffing box is a cylindrical space in the pump casing surrounding the shaft. Rings of packing material are placed in this space. Packing is material in the form of rings or strands that is placed in the stuffing box to form a seal to control the rate of leakage along the shaft. The packing rings are held in place by a gland. The gland is, in turn, held in place by studs with adjusting nuts. As the adjusting nuts are tightened, they move the gland in and compress the packing. This axial compression causes the packing to expand radially, forming a tight seal between the rotating shaft and the inside wall of the stuffing box.

The high-speed rotation of the shaft generates a significant amount of heat as it rubs against the packing rings. If no lubrication and cooling are provided to the packing, the temperature of the packing increases to the point where damage occurs to the packing, the pump shaft, and possibly the nearby pump bearing. Stuffing boxes are normally designed to allow a small amount of controlled leakage along the shaft to provide lubrication and cooling to the packing. Tightening and loosening the packing gland can adjust the leakage rate.

Lantern Ring

It is not always possible to use a standard stuffing box to seal the shaft of a centrifugal pump. The pump suction may be under a vacuum so that outward leakage is impossible or the fluid may be too hot to provide adequate cooling of the packing. These conditions require a modification to the standard stuffing box.

One method of adequately cooling the packing under these conditions is to include a lantern ring. A lantern ring is a perforated hollow ring located near the center of the packing box that receives relatively cool, clean liquid from either the discharge of the pump or from an external source and distributes the liquid uniformly around the shaft to provide lubrication and cooling. The fluid entering the lantern ring can cool the shaft and packing, lubricate the packing, or seal the joint between the shaft and packing against leakage of air into the pump in the event the pump suction pressure is less than that of the atmosphere.

Mechanical Seals

In some situations, packing material is not adequate for sealing the shaft. One common alternative method for sealing the shaft is with mechanical seals. Mechanical seals consist of two basic parts, a rotating element attached to the pump shaft and a stationary element attached to the pump casing. Each of these elements has a highly polished sealing surface. The polished faces of the rotating and stationary elements come into contact with each other to form a seal that prevents leakage along the shaft.

SUMMARY

The important information in this chapter is summarized below.

- Centrifugal pumps contain components with distinct purposes. The impeller contains rotating vanes that impart a radial and rotary motion to the liquid.
- The volute collects the liquid discharged from the impeller at high velocity and gradually causes a reduction in fluid velocity by increasing the flow area, converting the velocity head to a static head.

- A diffuser increases the efficiency of a centrifugal pump by allowing a more gradual expansion and less turbulent area for the liquid to slow as the flow area expands.
- Packing material provides a seal in the area where the pump shaft penetrates the pump casing.
- Wearing rings are replaceable rings that are attached to the impeller and/or the pump casing to allow a small running clearance between the impeller and pump casing without causing wear of the actual impeller or pump casing material.
- The lantern ring is inserted between rings of packing in the stuffing box to receive relatively cool, clean liquid and distribute the liquid uniformly around the shaft to provide lubrication and cooling to the packing.
- There are three indications that a centrifugal pump is cavitating:
  1. Noise
  2. Fluctuating discharge pressure and flow
  3. Fluctuating pump motor current
- Steps that can be taken to stop pump cavitation include:
  1. Increasing the pressure at the suction of the pump
  2. Reducing the temperature of the liquid being pumped
  3. Reducing head losses in the pump suction piping
  4. Reducing the flow rate through the pump
  5. Reducing the speed of the pump impeller
- Three effects of pump cavitation are:
  1. Degrading pump performance
  2. Excessive pump vibration
  3. Damage to pump impeller, bearing, wearing rings, and seals
- To avoid pump cavitation, the net positive suction head available must be greater than the net positive suction head required.
- Net positive suction head available is the difference between the pump suction pressure and the saturation pressure for the liquid being pumped.
- Cavitation is the process of the formation and subsequent collapse of vapor bubbles in a pump.
- Gas binding of a centrifugal pump is a condition in which the pump casing is filled with gases or vapors to the point where the impeller is no longer able to contact enough fluid to function correctly.
- Shutoff head is the maximum head that can be developed by a centrifugal pump operating at a set speed.
- Pump run-out is the maximum flow that can be developed by a centrifugal pump without damaging the pump.
- The greater the head against which a centrifugal pump operates, the lower the flow rate through the pump. The relationship between pump flow rate and head is illustrated by the characteristic curve for the pump.

- Centrifugal pumps are protected from deadheading by providing a recirculation from the pump discharge back to the supply source of the pump.
- Centrifugal pumps are protected from run-out by placing an orifice or throttle valve immediately downstream of the pump discharge.

POSITIVE-DISPLACEMENT PUMPS

A positive-displacement pump is one in which a definite volume of liquid is delivered for each cycle of pump operation. This volume is constant regardless of the resistance to flow offered by the system the pump is in, provided the capacity of the power unit driving the pump is not exceeded. The positive-displacement pump delivers liquid in separate volumes with no delivery in between, although a pump having several chambers may have an overlapping delivery among individual chambers, which minimizes this effect. The positive displacement pump differs from other types of pumps that deliver a continuous even flow for any given pump speed and discharge.

Positive-displacement pumps can be grouped into three basic categories based on their design and operation: reciprocating pumps, rotary pumps, and diaphragm pumps.

PRINCIPLES OF OPERATION

All positive-displacement pumps operate on the same basic principle. This principle can be most easily demonstrated by considering a reciprocating positive-displacement pump consisting of a single reciprocating piston in a cylinder with a single suction port and a single discharge port as shown in Figure 17.12.

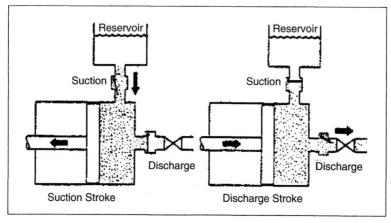

Figure 17.12 Reciprocating positive-displacement pump operation.

During the suction stroke, the piston moves to the left, causing the check valve in the suction line between the reservoir and the pump cylinder to open and admit water from the reservoir. During the discharge stroke, the piston moves to the right, seating the check valve in the suction line and opening the check valve in the discharge line. The volume of liquid moved by the pump in one cycle (one suction stroke and one discharge stroke) is equal to the change in the liquid volume of the cylinder as the piston moves from its farthest left position to its farthest right position.

RECIPROCATING PUMPS

Reciprocating positive-displacement pumps are generally categorized in four ways: direct-acting or indirect-acting; simplex or duplex; single-acting or double-acting; and power pumps.

Direct-Acting and Indirect-Acting

Some reciprocating pumps are powered by prime movers that also have reciprocating motion, such as a reciprocating pump powered by a reciprocating steam piston. The piston rod of the steam piston may be directly connected to the liquid piston of the pump or it may be indirectly connected with a beam or linkage. Direct-acting pumps have a plunger on the liquid (pump) end that is directly driven by the pump rod (also the piston rod or extension thereof) and carries the piston of the power end. Indirect-acting pumps are driven by means of a beam or linkage connected to and actuated by the power piston rod of a separate reciprocating engine.

Simplex and Duplex

A simplex pump, sometimes referred to as a *single* pump, is a pump having a single liquid (pump) cylinder. A duplex pump is the equivalent of two simplex pumps placed side by side on the same foundation.

The driving of the pistons of a duplex pump is arranged in such a manner that when one piston is on its up stroke, the other piston is on its down stroke and vice versa. This arrangement doubles the capacity of the duplex pump as compared with a simplex pump of comparable design.

Single-Acting and Double-Acting

A single-acting pump is one that takes a suction, filling the pump cylinder on the stroke in only one direction, called the *suction stroke*, and then forces the liquid out of the cylinder on the return stroke, called the *discharge stroke*. A double-acting pump is one that, as it fills one end of the liquid cylinder, is discharging liquid from the other end of the cylinder. On the return stroke, the end of the cylinder just emptied is filled, and the end just filled is emptied. One possible arrangement for single-acting and double-acting pumps is shown in Figure 17.13.

Power

Power pumps convert rotary motion to low-speed reciprocating motion by reduction gearing, a crankshaft, connecting rods, and crossheads. Plungers or pistons are driven by the crosshead drives. The liquid ends of the low-pressure, higher-capacity units use rod and piston construction similar to duplex double-acting steam pumps. The higher-pressure units are normally single-action plungers and usually employ three (triplex) plungers. Three or more plungers substantially reduce flow pulsations relative to simplex and even duplex pumps.

Power pumps typically have high efficiency and are capable of developing very high pressures. Either electric motors or turbines can drive them. They are relatively expensive pumps and can rarely be justified on the basis of efficiency over centrifugal pumps. However, they are frequently justified over steam reciprocating pumps in which continuous-duty service is needed because of the high steam requirements of direct-acting steam pumps.

In general, the effective flow rate of reciprocating pumps decreases as the viscosity of the fluid being pumped increases, because the speed of the pump must be reduced. In contrast to centrifugal pumps, the differential pressure generated by reciprocating pumps is independent of fluid density. It is dependent entirely on the amount of force exerted on the piston.

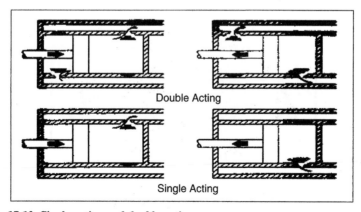

Double Acting

Single Acting

Figure 17.13 Single-acting and double-acting pumps.

ROTARY

Rotary pumps operate on the principle that a rotating vane, screw, or gear traps the liquid in the suction side of the pump casing and forces it to the discharge side of the casing. These pumps are essentially self-priming because of their capability of removing air from suction lines and producing a high suction lift. In pumps designed for systems requiring high suction lift and self-priming features, it is essential that all clearances between rotating parts, and between rotating and stationary parts, be kept to a minimum to reduce slippage. Slippage is leakage of fluid from the discharge of the pump back to its suction.

Because of the close clearances in rotary pumps, it is necessary to operate these pumps at relatively low speed to secure reliable operation and maintain pump capacity over an extended period of time. Otherwise, the erosive action caused by the high velocities of the liquid passing through the narrow clearance spaces would soon cause excessive wear and increased clearance, resulting in slippage.

There are many types of positive displacement rotary pumps, and they are normally grouped into three basic categories: gear pumps, screw pumps, and moving vane pumps.

Rotary Moving Vane

The rotary moving vane pump shown in Figure 17.14 is another type of positive-displacement pump used in pumping viscous fluids. The pump consists of a cylindrically bored housing with a suction inlet on one side and a discharge outlet on the other. A cylindrically shaped rotor, with a diameter smaller than

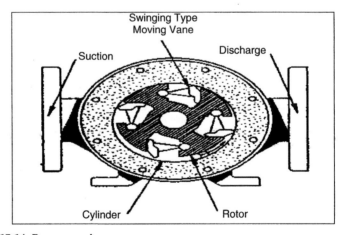

Figure 17.14 Rotary moving vane pump.

the cylinder, is driven about an axis place above the centerline of the cylinder. The clearance between rotor and cylinder at the top is small but increases at the bottom. The rotor carries vanes that move in and out as it rotates to maintain sealed space between the rotor and the cylinder wall. The vanes trap liquid on the suction side and carry it to the discharge side, where contraction of the space expels it through the discharge line. The vanes may swing on pivots, or they may slide in slots in the rotor.

Screw-Type, Positive-Displacement Rotary

There are many variations in the design of the screw-type, positive-displacement rotary pump. The primary differences consist of the number of intermeshing screws involved, the pitch of the screws, and the general direction of fluid flow. Two designs include a two-screw, low-pitch, double-flow pump and a three-screw, high-pitch, double-flow pump.

Two-Screw, Low-Pitch Screw

The two-screw, low-pitch screw pump consists of two screws that mesh with close clearances, mounted on two parallel shafts. One screw has a right-handed thread and the other screw has a left-handed thread. One shaft is the driving shaft and drives the other through a set of herringbone timing gears. The gears serve to maintain clearances between the screws as they turn and to promote quiet operation. The screws rotate in closely fitting duplex cylinders that have overlapping bores. All clearances are small, but there is no actual contact between the two screws or between the screws and the cylinder walls. The complete assembly and the usual path of flow are shown in Figure 17.15.

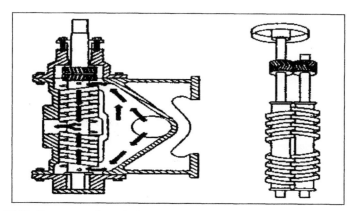

Figure 17.15 Two-screw, low-pitch screw pump.

Liquid is trapped at the outer end of each pair of screws. As the first space between the screw threads rotates away from the opposite screw, a one-turn, spiral-shaped quantity of liquid is enclosed when the end of the screw again meshes with the opposite screw. As the screw continues to rotate, the entrapped spiral turns of liquid slide along the cylinder toward the center discharge space while the next slug is being entrapped. Each screw functions similarly, and each pair of screws discharges an equal quantity of liquid in opposed streams toward the center, thus eliminating hydraulic thrust. The removal of liquid from the suction end by the screws produces a reduction in pressure, which draws liquid through the suction line.

Three-Screw, High-Pitch Screw

The three-screw, high-pitch screw pump shown in Figure 17.16 has many of the same elements as the two-screw, low-pitch screw pump, and their operations are similar. Three screws, oppositely threaded on each end, are employed. They rotate in a triple cylinder, the two outer bores of which overlap the center bore. The pitch of the screws is much higher than in the low-pitch screw pump; therefore the center screw, or power rotor, is used to drive the two outer idler rotors directly without external timing gears. Pedestal bearings at the base support the weight of the rotors and maintain their axial position as the liquid being pumped enters the suction opening, flows through passages around the rotor housing, and flows through the screws from each end, in opposed streams, toward the center discharge. This eliminates unbalanced hydraulic thrust. The screw pump is used for pumping viscous fluids, usually lubricating, hydraulic, or fuel oil.

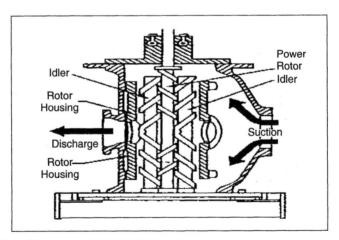

Figure 17.16 Three-screw, high-pitch screw pump.

Diaphragm or Positive Displacement

Diaphragm pumps are also classified as positive-displacement pumps because the diaphragm acts as a limited-displacement piston. The pump will function when a diaphragm is forced into reciprocating motion by mechanical linkage, compressed air, or fluid from a pulsating external source. The pump construction eliminates any contact between the liquid being pumped and the source of energy. This eliminates the possibility of leakage, which is important when handling toxic or very expensive liquids. Disadvantages include limited head and capacity range, and the necessity of check valves in the suction and discharge nozzles. An example of a diaphragm pump is shown in Figure 17.17.

Characteristic Curve

Positive-displacement pumps deliver a definite volume of liquid for each cycle of pump operation. Therefore, the only factor that affects flow rate in an ideal positive-displacement is the speed at which it operates. The flow resistance of the system in which the pump is operating will not affect the flow rate through the pump. Figure 17.18 shows the characteristic curve for a positive-displacement pump.

The dashed line in Figure 17.18 shows actual positive-displacement pump performance. This line reflects the fact that as the discharge pressure of the pump increases, some amount of liquid will leak from the discharge of the pump back to the pump suction, reducing the effective flow rate of the pump. The rate at which liquid leaks from the pump discharge to its suction is called *slippage*.

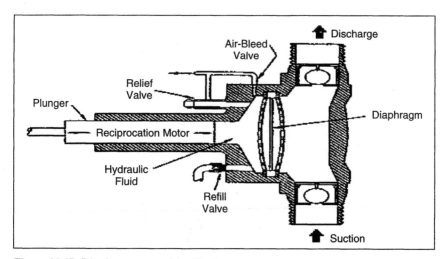

Figure 17.17 Diaphragm or positive-displacement pump.

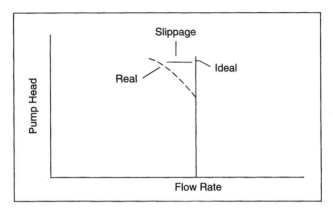

Figure 17.18 Positive-displacement pump characteristic curve.

Protection

Positive-displacement pumps are normally fitted with relief valves on the upstream side of their discharge valves to protect the pump and its discharge piping from over-pressurization. Positive-displacement pumps will discharge at the pressure required by the system they are supplying. The relief valve prevents system and pump damage if the pump discharge valve is shut during pump operation or if any other occurrence such as a clogged strainer blocks system flow.

GEAR PUMPS

Simple Gear Pumps

There are several variations of gear pumps. The simple gear pump shown in Figure 17.19 consists of two spur gears meshing together and revolving in opposite directions within a casing. Only a few thousandths of an inch clearance

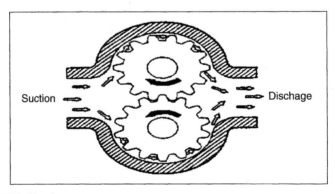

Figure 17.19 Simple gear pump.

exists between the case and the gear faces and teeth extremities. Any liquid that fills the space bounded by two successive gear teeth and the case must follow along with the teeth as they revolve. When the gear teeth mesh with the teeth of the other gear, the space between the teeth is reduced, and the entrapped liquid is forced out the pump discharge pipe. As the gears revolve and the teeth disengage, the space again opens on the suction side of the pump, trapping new quantities of liquid and carrying it around the pump case to the discharge. As liquid is carried away from the suction side, a lower pressure is created, which draws liquid in through the suction line.

With the large number of teeth usually employed on the gears, the discharge is relatively smooth and continuous, with small quantities of liquid being delivered to the discharge line in rapid succession. If designed with fewer teeth, the space between the teeth is greater and the capacity increases for a given speed; however, the tendency toward a pulsating discharge increases. In all simple gear pumps, power is applied to the shaft of one of the gears, which transmits power to the driven gear through their meshing teeth.

There are no valves in the gear pump to cause friction losses as in the recipro-cating pump. The high impeller velocities, with resultant friction losses, are not required as in the centrifugal pump. Therefore the gear pump is well suited for handling viscous fluids such as fuel and lubricating oils.

Other Gear Pumps

There are two types of gears used in gear pumps in addition to the simple spur gear. One type is the helical gear. A helix is the curve produced when a straight line moves up or down the surface of a cylinder. The other type is the herringbone gear. A herringbone gear is composed of two helixes spiraling in different directions from the center of the gear. Spur, helical, and herringbone gears are shown in Figure 17.20.

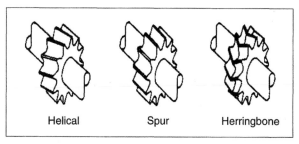

Figure 17.20 Types of gears used in pumps.

The helical gear pump has advantages over the simple spur gear. In a spur gear, the entire length of the gear tooth engages at the same time. In a helical gear, the point of engagement moves along the length of the gear tooth as the gear rotates. This makes the helical gear operate with a steadier discharge pressure and fewer pulsations than a spur gear pump.

The herringbone gear pump is also a modification of the simple gear pump. Its principal difference in operation from the simple gear pump is that the pointed center section of the space between two teeth begins discharging before the divergent outer ends of the preceding space complete discharging. This overlapping tends to provide a steadier discharge pressure. The power transmission from the driving to the driven gear is also smoother and quieter.

Lobe-Type Pump

The lobe-type pump shown in Figure 17.21 is another variation of the simple gear pump. It is considered as a simple gear pump having only two or three teeth per rotor; otherwise, its operation or the explanation of the function of its parts is no different. Some designs of lobe pumps are fitted with replaceable gibs, that is, thin plates carried in grooves at the extremity of each lobe where they make contact with the casing. The gibs promote tightness and absorb radial wear.

Summary

The important information in this chapter is summarized below.

- The flow delivered by a centrifugal pump during one revolution of the impeller depends on the head against which the pump is operating. The positive-displacement pump delivers a fixed volume of fluid for each

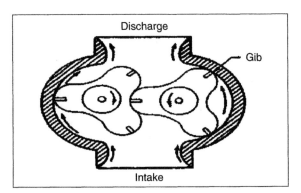

Figure 17.21 Lobe-type pump.

cycle of pump operation regardless of the head against which the pump is operating.

- Positive displacement pumps may be classified in the following ways:
    1. Reciprocating piston pump
    2. Gear-type rotary pump
    3. Lobe-type rotary pump
    4. Screw-type rotary pump
    5. Moving vane pump
    6. Diaphragm pump
- As the viscosity of a liquid increases, the maximum speed at which a reciprocating positive displacement pump can properly operate decreases. Therefore, as viscosity increases, the maximum flow rate through the pump decreases.
- The characteristic curve for a positive displacement pump can properly operate decreases. Therefore, as viscosity increases, the maximum flow rate through the pump decreases.
- Slippage is the rate at which liquid leaks from the discharge of the pump back to the pump suction.
- Positive displacement pumps are protected from over-pressurization by a relief valve on the upstream side of the pump discharge valve.

CAVITATION

Many centrifugal pumps are designed in a manner that allows the pump to operate continuously for months or even years. These centrifugal pumps often rely on the liquid that they are pumping to provide cooling and lubrication to the pump bearings and other internal components of the pump. If flow through the pump is stopped while the pump is still operating, the pump will no longer be adequately cooled and the pump can quickly become damaged. Pump damage can also result from pumping a liquid that is close to saturated conditions. This phenomenon, referred to as *cavitation*, is discussed further in Chapter 5 (Preventing Cavitation). Most centrifugal pumps are not designed to withstand cavitation.

The flow area at the eye of the impeller is usually smaller than either the flow area of the pump suction piping or the flow area through the impeller vanes. When the liquid being pumped enters the eye of a centrifugal pump, the decrease in flow area results in an increase in flow velocity accompanied by a decrease in pressure. The greater the pump flow rate, the greater the pressure drop between the pump suction and the eye of the impeller. If the pressure drop is large enough, or if the temperature is high enough, the pressure drop may be sufficient to cause the liquid to flash to vapor when the local pressure falls below the saturation pressure for the fluid being pumped. Any vapor bubbles formed by the pressure drop at the eye of the impeller are swept along the impeller vanes by the flow of the fluid. When the bubbles enter a region in which local pressure is greater than saturation pressure farther out the impeller vane, the vapor bubbles abruptly collapse. This process of the formation and subsequent collapse of vapor bubbles in a pump is called *cavitation.*

Cavitation in a centrifugal pump has a significant effect on performance. It degrades the performance of a pump, resulting in a degraded, fluctuating flow rate and discharge pressure. Cavitation can also be destructive to pump internals. The formation and collapse of the vapor bubble can create small pits on the impeller vanes. Each individual pit is microscopic in size, but the cumulative effect of millions of these pits formed over a period of hours or days can literally destroy a pump impeller. Cavitation can also cause excessive pump vibration, which could damage pump bearings, wearing rings, and seals.

A small number of centrifugal pumps are designed to operate under conditions in which cavitation is unavoidable. These pumps must be specially designed and maintained to withstand the small amount of cavitation that occurs during their operation.

Noise is one of the indications that a centrifugal pump is cavitating. A cavitating pump can sound like a can of marbles being shaken. Other indications that can be observed from a remote operating station are fluctuating discharge pressure, flow rate, and pump motor current.

RECIRCULATION

When the discharge flow of a centrifugal pump is throttled by closing the discharge valve slightly, or by installing an orifice plate, the fluid flow through the pump is altered from its original design. This reduces the fluid's velocity as it exits the tips of the impeller vanes; therefore the fluid does not flow as smoothly into the volute and discharge nozzle. This causes the fluid to impinge on the "cutwater" and creates a vibration at a frequency equal to the vane pass x rpm. The resulting amplitude quite often exceeds alert setpoint values, particularly when accompanied by resonance.

Random low-amplitude, wide-frequency vibration is often associated with vane pass frequency, resulting in vibrations similar to cavitation and turbulence, but is usually found at lower frequencies. This can lead to misdiagnosis. Many pump impellers show metal reduction and pitting on the general area at the exit tips of the vanes. This has often been misdiagnosed as cavitation.

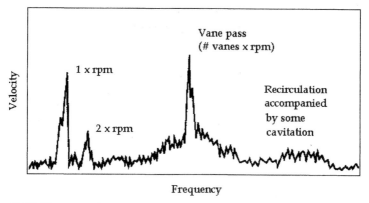

Figure 17.22 Vane pass frequency.

It is very important to note that recirculation is found to happen on the discharge side of the pump, whereas cavitation is found to happen on the suction side of the pump.

To prevent recirculation in pumps, pumps should be operated close to their operational rated capacity, and excessive throttling should be avoided.

When a permanent reduction in capacity is desired, the outside diameter of the pump impeller can be reduced slightly to increase the gap between the impeller tips and the cutwater.

NET POSITIVE SUCTION HEAD

To avoid cavitation in centrifugal pumps, the pressure of the fluid at all points within the pump must remain above saturation pressure. The quantity used to determine whether the pressure of the liquid being pumped is adequate to avoid cavitation is the net positive suction head (NPSH). The net positive suction head available ($NPSH_A$) is the difference between the pressure at the suction of the pump and the saturation pressure for the liquid being pumped. The net positive suction head required ($NPSH_R$) is the minimum net positive suction head necessary to avoid cavitation.

The condition that must exist to avoid cavitation is that the net positive suction head available must be greater than or equal to the net positive suction head required. This requirement can be stated mathematically as shown below.

$$NPSH_A \geq NPSH_R$$

A formula for NPSH$_A$ can be stated as the following equation:

$$NPSH_A = P_{suction} - P_{saturation}$$

When a centrifugal pump is taking suction from a tank or other reservoir, the pressure at the suction of the pump is the sum of the absolute pressure at the surface of the liquid in the tank plus the pressure caused by the elevation difference between the surface of liquid in the tank and the pump suction, less the head losses caused by friction in the suction line from the tank to the pump.

$$NPSH_A = P_a = P_{st} - h_f - P_{sat}$$

where

$\quad$ NPSH$_A$ = Net positive suction head available

$\quad\quad$ P_a = Absolute pressure on the surface of the liquid

$\quad\quad$ P_{st} = Pressure caused by elevation between liquid surface and pump suction

$\quad\quad$ h_f = Head losses in the pump suction piping

$\quad\quad$ P_{sat} = Saturation pressure of the liquid being pumped

PREVENTING CAVITATION

If a centrifugal pump is cavitating, several changes in the system design or operation may be necessary to increase the NPSH$_A$ above the NPSH$_R$ and stop the cavitation. One method for increasing the NPSH$_A$ is to increase the pressure at the suction of the pump. If a pump is taking suction from an enclosed tank, either raising the level of the liquid in the tank or increasing the pressure in the gas space above the liquid increases suction pressure.

It is also possible to increase the NPSH$_A$ by decreasing the temperature of the liquid being pumped. Decreasing the temperature of the liquid decreases the saturation pressure, causing NPSH$_A$ to increase.

If the head losses in the pump suction piping can be reduced, the NPSH$_A$ will be increased. Various methods for reducing head losses include increasing the pipe diameter; reducing the number of elbows, valves, and fittings in the pipe; and decreasing the length of the pipe.

It may also be possible to stop cavitation by reducing the NPSH$_R$ for the pump. The NPSH$_R$ is not a constant for a given pump under all conditions but depends on certain factors. Typically, the $NPSH_R$ of a pump increases significantly as flow rate through the pump increases. Therefore, reducing the flow rate through a pump by throttling a discharge valve decreases $NPSH_R$. $NPSH_R$ is also dependent on pump speed. The faster the impeller of a pump rotates, the greater the $NPSH_R$. Therefore, if the speed of a variable-speed centrifugal pump is reduced, the $NPSH_R$ of the pump decreases.

The net positive suction head required to prevent cavitation is determined through testing by the pump manufacturer and depends on factors including type of impeller inlet, impeller design, pump flow rate, impeller rotational speed, and the type of liquid being pumped. The manufacturer typically supplies curves of $NPSH_R$ as a function of pump flow rate for a particular liquid (usually water) in the vendor manual for the pump.

TROUBLESHOOTING

Design, installation, and operation are the dominant factors that affect a pump's mode of failure. This section identifies common failures for centrifugal and positive-displacement pumps.

CENTRIFUGAL

Centrifugal pumps are especially sensitive to (1) variations in liquid condition (i.e., viscosity, specific gravity, and temperature); (2) suction variations, such as pressure and availability of a continuous volume of fluid; and (3) variations in demand. Table 17.1 lists common failure modes for centrifugal pumps and their causes.

Mechanical failures may occur for a number of reasons. Some are induced by cavitation, hydraulic instability, or other system-related problems. Others are the direct result of improper maintenance. Maintenance-related problems include improper lubrication, misalignment, imbalance, seal leakage, and a variety of others that periodically affect machine reliability.

Cavitation

Cavitation in a centrifugal pump, which has a significant, negative effect on performance, is the most common failure mode. Cavitation not only degrades a pump's performance, it also greatly accelerates the wear rate of its internal components.

Table 17.1 Common Failure Modes of Centrifugal Pumps

THE CAUSES	Insufficient Discharge Pressure	Intermittent Operation	Insufficient Capacity	No Liquid Delivery	High Bearing Temperatures	Short Bearing Life	Short Mechanical Seal Life	High Vibration	High Noise Levels	Power Demand Excessive	Motor Trips	Elevated Motor Temperature	Elevated Liquid Temperature
Bent Shaft					●	●	●	●		●			
Casing Distorted from Excessive Pipe Strain					●	●	●	●		●		●	
Cavitation	●	●	●	●	●		●	●	●				●
Clogged Impeller	●		●	●				●		●			
Driver Imbalance					●	●	●						
Electrical Problems (Driver)					●	●	●	●		●	●	●	
Entrained Air (Suction or Seal Leaks)	●	●	●					●	●			●	
Hydraulic Instability					●	●	●	●	●				
Impeller Installed Backward (Double-Suction Only)	●		●							●			
Improper Mechanical Seal							●						
Inlet Strainer Partially Clogged	●		●					●	●				●
Insufficient Flow Through Pump													●
Insufficient Suction Pressure (NPSH)	●	●	●	●				●	●				
Insufficient Suction Volume	●	●	●	●	●			●	●				●
Internal Wear	●		●					●		●			
Leakage in Piping, Valves, Vessels	●		●	●									
Mechanical Defects, Worn, Rusted, Defective Bearings					●		●			●			
Misalignment					●	●	●	●		●		●	
Misalignment (Pump and Driver)								●		●	●		●
Mismatched Pumps In Series	●		●			●		●		●			
Non-condensables in Liquid	●	●	●					●	●			●	
Obstructions in Lines or Pump Housing	●		●	●				●				●	●
Rotor Imbalance					●	●	●						
Specific Gravity Too High	●									●		●	
Speed Too High										●	●		
Speed Too Low	●		●	●								●	
Total System Head Higher Than Design	●	●	●	●	●		●					●	●
Total System Head Lower Than Design						●			●	●	●	●	●
Unsuitable Pumps in Parallel Operation	●		●	●				●	●		●		●
Viscosity Too High	●		●							●		●	
Wrong Rotation	●			●						●		●	

Source: Integrated Systems, Inc.

Causes

There are three causes of cavitation in centrifugal pumps: change of phase, entrained air or gas, and turbulent flow.

Change of Phase The formation or collapse of vapor bubbles in either the suction piping or inside the pump is one cause of cavitation. This failure mode normally occurs in applications such as boiler feed in which the incoming liquid is at a temperature near its saturation point. In this situation, a slight change in suction pressure can cause the liquid to flash into its gaseous state. In the boiler-feed example, the water flashes into steam. The reverse process also can occur. A slight increase in suction pressure can force the entrained vapor to change phase to a liquid.

Cavitation caused by phase change seriously damages the pump's internal components. Visual evidence of operation with phase-change cavitation is an impeller surface finish like an orange peel. Prolonged operation causes small pits or holes on both the impeller shroud and vanes.

Entrained Air/Gas Pumps are designed to handle gas-free liquids. If a centrifugal pump's suction supply contains any appreciable quantity of gas, the pump will cavitate. In the example of cavitation caused by entrainment, the liquid is reasonably stable, unlike with the change of phase described in the preceding section. Nevertheless, the entrained gas has a negative effect on pump performance. While this form of cavitation does not seriously affect the pump's internal components, it severely restricts its output and efficiency.

The primary causes of cavitation that is due to entrained gas include two-phase suction supply, inadequate available net positive suction head ($NPSH_A$), and leakage in the suction-supply system. In some applications, the incoming liquid may contain moderate to high concentrations of air or gas. This may result from aeration or mixing of the liquid prior to reaching the pump or inadequate liquid levels in the supply reservoir. Regardless of the reason, the pump is forced to handle two-phase flow, which was not intended in its design.

Turbulent Flow The effects of turbulent flow (not a true form of cavitation) on pump performance are almost identical to those described for entrained air or gas in the preceding section. Pumps are not designed to handle incoming liquids that do not have stable, laminar flow patterns. Therefore, if the flow is unstable or turbulent, the symptoms are the same as for cavitation.

Symptoms Noise (e.g., like a can of marbles being shaken) is one indication that a centrifugal pump is cavitating. Other indications are fluctuations of the pressure gauges, flow rate, and motor current, as well as changes in the vibration profile.

How to Eliminate Several design or operational changes may be necessary to stop centrifugal-pump cavitation. Increasing the available net positive suction head ($NPSH_A$) above that required ($NPSH_R$) is one way to stop it. The NPSH required to prevent cavitation is determined through testing by the pump manufacturer. It depends on several factors, including type of impeller inlet, impeller design, impeller rotational speed, pump flow rate, and the type of liquid being pumped. The manufacturer typically supplies curves of $NPSH_R$ as a function of flow rate for a particular liquid (usually water) in the pump's manual.

One way to increase the $NPSH_A$ is to increase the pump's suction pressure. If a pump is fed from an enclosed tank, either raising the level of the liquid in the tank or increasing the pressure in the gas space above the liquid can increase suction pressure.

It also is possible to increase the $NPSH_A$ by decreasing the temperature of the liquid being pumped. This decreases the saturation pressure, which increases $NPSH_A$.

If the head losses in the suction piping can be reduced, the $NPSH_A$ will be increased. Methods for reducing head losses include increasing the pipe diameter; reducing the number of elbows, valves, and fittings in the pipe; and decreasing the pipe length.

It also may be possible to stop cavitation by reducing the pump's $NPSH_R$, which is not a constant for a given pump under all conditions. Typically, the $NPSH_R$ increases significantly as the pump's flow rate increases. Therefore, reducing the flow rate by throttling a discharge valve decreases $NPSH_R$. In addition to flow rate, $NPSH_R$ depends on pump speed. The faster the pump's impeller rotates, the greater the $NPSH_R$. Therefore, if the speed of a variable-speed centrifugal pump is reduced, the $NPSH_R$ of the pump is decreased.

Variations in Total System Head

Centrifugal-pump performance follows its hydraulic curve (i.e., head versus flow rate). Therefore any variation in the total backpressure of the system causes a change in the pump's flow or output. Because pumps are designed to operate at their BEP, they become more and more unstable as they are forced to operate at any other point because of changes in total system pressure, or head (TSH). This instability has a direct impact on centrifugal-pump performance, reliability, operating costs, and required maintenance.

Symptoms of Changed Conditions

The symptoms of failure caused by variations in TSH include changes in motor speed and flow rate.

Motor Speed The brake horsepower of the motor that drives a pump is load dependent. As the pump's operating point deviates from BEP, the amount of horsepower required also changes. This causes a change in the pump's rotating speed, which either increases or decreases depending on the amount of work that the pump must perform.

Flow Rate The volume of liquid delivered by the pump varies with changes in TSH. An increase in the total system backpressure results in decreased flow, while a backpressure reduction increases the pump's output.

Correcting Problems The best solution to problems caused by TSH variations is to prevent the variations. While it is not possible to completely eliminate them, the operating practices for centrifugal pumps should limit operation to an acceptable range of system demand for flow and pressure. If system demand exceeds the pump's capabilities, it may be necessary to change the pump, the system requirements, or both. In many applications, the pump is either too small or too large. In these instances, it is necessary to replace the pump with one that is properly sized.

For the application in which the TSH is too low and the pump is operating in run-out condition (i.e., maximum flow and minimum discharge pressure), the system demand can be corrected by restricting the discharge flow of the pump. This approach, called *false head*, changes the system's head by partially closing a discharge valve to increase the backpressure on the pump. Because the pump must follow its hydraulic curve, this forces the pump's performance back towards its BEP.

When the TSH is too great, there are two options: replace the pump or lower the system's backpressure by eliminating line resistance caused by elbows, extra valves, etc.

Positive Displacement

Positive-displacement pumps are more tolerant of variations in system demands and pressures than centrifugal pumps. However, they are still subject to a variety of common failure modes caused directly or indirectly by the process.

Rotary Type

Rotary-type, positive-displacement pumps share many common failure modes with centrifugal pumps. Both types of pumps are subject to process-induced failures caused by demands that exceed the pump's capabilities. Process-induced failures also are caused by operating methods that either result in radical changes in their operating envelope or instability in the process system.

Table 17.2 lists common failure modes for rotary-type, positive-displacement pumps. The most common failure modes of these pumps are generally attributed to problems with the suction supply. They must have a constant volume of clean liquid to function properly.

RECIPROCATING

Table 17.3 lists the common failure modes for reciprocating-type, positive-displacement pumps. Reciprocating pumps can generally withstand more abuse and variations in system demand than any other type. However, they must have a consistent supply of relatively clean liquid to function properly.

Table 17.2 Common Failure Modes of Rotary-Type, Positive-Displacement Pumps

THE CAUSES	No Liquid Delivery	Insufficient Discharge Pressure	Insufficient Capacity	Starts, But Loses Prime	Excessive Wear	Excessive Heat	Excessive Vibration and Noise	Excessive Power Demand	Motor Trips	Elevated Motor Temperature	Elevated Liquid Temperature
Air Leakage into Suction Piping or Shaft Seal		●	●				●			●	
Excessive Discharge Pressure			●		●		●	●	●		●
Excessive Suction Liquid Temperatures			●	●							
Insufficient Liquid Supply		●	●	●	●		●		●		
Internal Component Wear	●	●	●				●				
Liquid More Viscous Than Design								●	●	●	●
Liquid Vaporizing in Suction Line		●	●	●			●				●
Misaligned Coupling, Belt Drive, Chain Drive					●	●	●	●		●	
Motor or Driver Failure	●										
Pipe Strain on Pump Casing					●	●	●	●		●	
Pump Running Dry	●	●			●	●	●				
Relief Valve Stuck Open or Set Wrong		●	●								
Rotating Element Binding					●	●	●	●	●	●	
Solids or Dirt in Liquid					●						
Speed Too Low		●	●							●	
Suction Filter or Strainer Clogged	●	●	●				●			●	
Suction Piping Not Immersed in Liquid	●	●		●							
Wrong Direction of Rotation	●	●								●	

Source: Integrated Systems, Inc.

Table 17.3 Common Failure Modes of Reciprocating Positive-Displacement Pumps

THE CAUSES	No Liquid Delivery	Insufficient Capacity	Short Packing Life	Excessive Wear Liquid End	Excessive Wear Power End	Excessive Heat Power End	Excessive Vibration and Noise	Persistent Knocking	Motor Trips
Abrasives or Corrosives in Liquid			●	●					
Broken Valve Springs		●		●			●		
Cylinders Not Filling		●	●	●			●		
Drive-train Problems							●		●
Excessive Suction Lift	●	●							
Gear Drive Problem							●	●	●
Improper Packing Selection			●						
Inadequate Lubrication						●	●		●
Liquid Entry into Power End of Pump						●			
Loose Cross-head Pin or Crank Pin								●	
Loose Piston or Rod								●	
Low Volumetric Efficiency		●	●						
Misalignment of Rod or Packing			●						●
Non-condensables (Air) in Liquid	●	●	●				●		●
Not Enough Suction Pressure	●	●							
Obstructions in Lines	●						●		●
One or More Cylinders Not Operating		●							
Other Mechanical Problems: Wear, Rusted, etc.					●	●	●	●	
Overloading						●			●
Pump Speed Incorrect		●				●			
Pump Valve(s) Stuck Open		●							
Relief or Bypass Valve(s) Leaking		●							
Scored Rod or Plunger		●							●
Supply Tank Empty	●								
Worn Cross-head or Guides			●			●			
Worn Valves, Seats, Liners, Rods, or Plungers	●	●		●					

Source: Integrated Systems, Inc.

The weak links in the reciprocating pump's design are the inlet and discharge valves used to control pumping action. These valves are the most frequent source of failure. In most cases, valve failure is caused by fatigue. The only positive way to prevent or minimize these failures is to ensure that proper maintenance is performed regularly on these components. It is important to follow the manufacturer's recommendations for valve maintenance and replacement.

Because of the close tolerances between the pistons and the cylinder walls, reciprocating pumps cannot tolerate contaminated liquid in their suction-supply system. Many of the failure modes associated with this type of pump are caused by contamination (e.g., dirt, grit, and other solids) that enters the suction side of the pump. This problem can be prevented by the use of well-maintained inlet strainers or filters.

18

STEAM TRAPS

Steam-supply systems are commonly used in industrial facilities as a general heat source and as a heat source in pipe and vessel tracing lines used to prevent freeze-up in non-flow situations. Inherent with the use of steam is the problem of condensation and the accumulation of non-condensable gases in the system.

Steam traps must be used in these systems to automatically purge condensate and non-condensable gases, such as air, from the steam system. However, a steam trap should never discharge live steam. Such discharges are dangerous as well as costly.

CONFIGURATION

Five major types of steam traps are commonly used in industrial applications: inverted bucket, float and thermostatic, thermodynamic, bimetallic, and thermostatic. Each of the five major types of steam trap uses a different method to determine when and how to purge the system. As a result, each has a different configuration.

Inverted Bucket

The inverted-bucket trap, which is shown in Figure 18.1, is a mechanically actuated steam trap that uses an upside down, or inverted, bucket as a float. The bucket is connected to the outlet valve through a mechanical linkage. The bucket sinks when condensate fills the steam trap, which opens the outlet valve and drains the bucket. It floats when steam enters the trap and closes the valve.

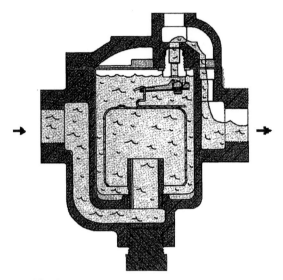

Figure 18.1 Inverted-bucket trap.

As a group, inverted-bucket traps can handle a wide range of steam pressures and condensate capacities. They are an economical solution for low- to medium-pressure and medium-capacity applications such as plant heating and light processes. When used for higher-pressure and higher-capacity applications, these traps become large, expensive, and difficult to handle.

Each specific steam trap has a finite, relatively narrow range that it can handle effectively. For example, an inverted-bucket trap designed for up to 15-psi service will fail to operate at pressures above that value. An inverted-bucket trap designed for 125-psi service will operate at lower pressures, but its capacity is so diminished that it may back up the system with unvented condensate. Therefore it is critical to select a steam trap designed to handle the application's pressure, capacity, and size requirements.

Float-and-Thermostatic

The float-and-thermostatic trap shown in Figure 18.2 is a hybrid. A float similar to that found in a toilet tank operates the valve. As condensate collects in the trap, it lifts the float and opens the discharge or purge valve. This design opens the discharge only as much as necessary. Once the built-in thermostatic element purges non-condensable gases, it closes tightly when steam enters the trap. The advantage of this type of trap is that it drains condensate continuously.

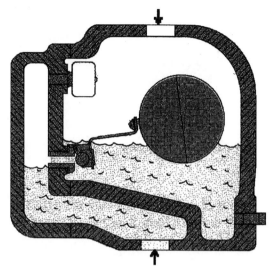

Figure 18.2 Float-and-thermostatic trap.

Like the inverted-bucket trap, float-and-thermostatic traps as a group handle a wide range of steam pressures and condensate loads. However, each individual trap has a very narrow range of pressures and capacities. This makes it critical to select a trap that can handle the specific pressure, capacity, and size requirements of the system.

The key advantage of float-and-thermostatic traps is their ability for quick steam-system startup because they continuously purge the system of air and other non-condensable gases. One disadvantage is the sensitivity of the float ball to damage by hydraulic hammer.

Float-and-thermostatic traps provide an economical solution for lighter conden-sate loads and lower pressures. However, when the pressure and capacity requirements increase, the physical size of the unit increases and its cost rises. It also becomes more difficult to handle.

Thermodynamic, or Disk-Type

Thermodynamic, or disk-type, steam traps use a flat disk that moves between a cap and seat (see Figure 18.3). On startup, condensate flow raises the disk and opens the discharge port. Steam or very hot condensate entering the trap seats the disk. It remains seated, closing the discharge port, as long as pressure is maintained above it. Heat radiates out through the cap, thus diminishing the pressure over the disk, opening the trap to discharge condensate.

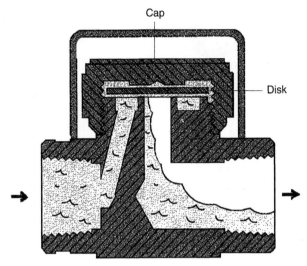

Figure 18.3 Thermodynamic steam trap.

Wear and dirt are particular problems with a disk-type trap. Because of the large, flat seating surfaces, any particulate contamination such as dirt or sand will lodge between the disk and the valve seat. This prevents the valve from sealing and permits live steam to flow through the discharge port. If pressure is not maintained above the disk, the trap will cycle frequently. This wastes steam and can cause the device to fail prematurely.

The key advantage of these traps is that one trap can handle a complete range of pressures. In addition, they are relatively compact for the amount of condensate they discharge. The chief disadvantage is difficulty in handling air and other non-condensable gases.

Bimetallic

A bimetallic steam trap, which is shown in Figure 18.4, operates on the same principle as a residential-heating thermostat. A bimetallic strip, or wafer, connected to a valve disk bends or distorts when subjected to a change in temperature. When properly calibrated, the disk closes tightly against a seat when steam is present and opens when condensate, air, and other gases are present.

Two key advantages of bimetallic traps are (1) compact size relative to their condensate load-handling capabilities and (2) immunity to hydraulic-hammer damage.

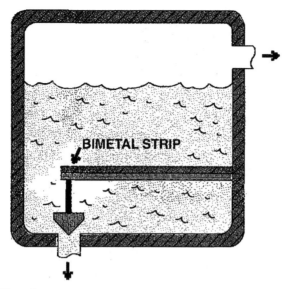

Figure 18.4 Bimetal trap.

Their biggest disadvantage is the need for constant adjustment or calibration, which is usually done at the factory for the intended steam operating pressure. If the trap is used at a lower pressure, it may discharge live steam. If used at a higher pressure, condensate may back up into the steam system.

Thermostatic or Thermal-Element

Thermostatic or thermal-element traps are thermally actuated by using an assembly constructed of high-strength, corrosion-resistant stainless steel plates that are seam-welded together. Figure 18.5 shows this type of trap.

Upon startup, the thermal element is positioned to open the valve and purge condensate, air, and other gases. As the system warms up, heat generates pressure in the thermal element, causing it to expand and throttle the flow of hot condensate through the discharge valve. The steam that follows the hot condensate into the trap expands the thermal element with great force, which causes the trap to close. Condensate that enters the trap during system operation cools the element. As the thermal element cools, it lifts the valve off the seat and allows condensate to discharge quickly.

Thermal elements can be designed to operate at any steam temperature. In steam-tracing applications, it may be desirable to allow controlled amounts of condensate to back up in the lines to extract more heat from the condensate.

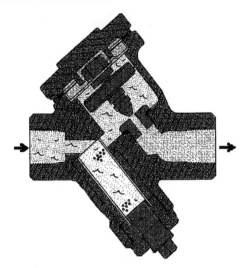

Figure 18.5 Thermostatic trap.

In other applications, any hint of condensate in the system is undesirable. The thermostatic trap can handle either of these conditions, but the thermal element must be properly selected to accommodate the specific temperature range of the application.

Thermostatic traps are compact, and a given trap operates over a wide range of pressures and capacities. However, they are not recommended for condensate loads over 15,000 pounds per hour.

PERFORMANCE

When properly selected, installed, and maintained, steam traps are relatively trouble-free and highly efficient. The critical factors that affect efficiency include capacity and pressure ratings, steam quality, mechanical damage, and calibration.

Capacity Rating

Each type and size of steam trap has a specified capacity for the amount of condensate and non-compressible gas that it can handle. Care must be taken to ensure that the proper steam trap is selected to meet the application's capacity needs.

Pressure Rating

As discussed previously, each type of steam trap has a range of steam pressures that it can effectively handle. Therefore, each application must be carefully evaluated to determine the normal and maximum pressures that will be generated by the steam system. Traps must be selected for the worst-case scenario.

Steam Quality

Steam quality determines the amount of condensate to be handled by the steam trap. In addition to an increased volume of condensate, poor steam quality may increase the amount of particulate matter present in the condensate. High concentrations of solids directly affect the performance of steam traps. If particulate matter is trapped between the purge valve and its seat, the steam trap may not properly shut off the discharge port. This will result in live steam being continuously exhausted through the trap.

Mechanical Damage

Inverted-bucket and float-type steam traps are highly susceptible to mechanical damage. If the level arms or mechanical linkages are damaged or distorted, the trap cannot operate properly. Regular inspection and maintenance of these types of traps are essential.

Calibration

Steam traps, such as the bimetallic type, must be periodically recalibrated to ensure proper operation. All steam traps should be adjusted on a regular schedule.

INSTALLATION

Installation of steam traps is relatively straightforward. As long as they are properly sized, the only installation imperative is that they are plumb. If the trap is tilted or cocked, the bucket, float, or thermal valve will not operate properly. In addition, a non-plumb installation may prevent the condensate chamber from fully discharging accumulated liquids.

OPERATING METHODS

Steam traps are designed for a relatively constant volume, pressure, and condensate load. Operating practices should attempt to maintain these parameters as

Table 18.1   Common Failure Modes of Steam Traps

THE CAUSES	Trap Will Not Discharge	Will Not Shut Off	Continuously Blows Steam	Capacity Suddenly Falls Off	Condensate Will Not Drain	Not Enough Steam Heat	Traps Freeze in Winter	Back Flow in Return Line
Backpressure Too High				●				
Boiler Foaming or Priming		●				●		
Boiler Gauge Reads Low	●							
Bypass Open or Leaking	●		●					
Condensate Load Greater Than Design		●						
Condensate Short-circuits					●			
Defective Thermostatic Elements						●		
Dirt or Scale In Trap			●		●			
Discharge Line Has Long Horizontal Runs							●	
Flashing in Return Main				●				●
High-pressure Traps Discharge into Low-pressure Return								●
Incorrect Fittings or Connectors				●				●
Internal Parts of Trap Broken or Damaged	●	●	●		●			
Internal Parts of Trap Plugged	●				●			
Kettles or Other Units Increasing Condensate Load		●						
Leaky Steam Coils		●						
No Cooling Leg Ahead of Thermostatic Trap						●		●
Open Bypass or Vent in Return Line				●				
Pressure Regulator Out of Order	●							
Process Load Greater Than Design		●						
Plugged Return Lines				●				
Plugged Strainer, Valve, or Fitting Ahead of Trap	●							
Scored or Out-of-Round Valve Seat in Trap						●		
Steam Pressure Too High	●							
System is Air-bound					●			
Trap and Piping Not Insulated							●	
Trap Below Return Main				●				●
Trap Blowing Steam into Return				●				
Trap Inlet Pressure Too Low				●	●			
Trap Too Small for Load		●						

Source: Integrated Systems, Inc.

much as possible. Actual operating practices are determined by the process system rather than the trap selected for a specific system.

The operator should periodically inspect them to ensure proper operation. Special attention should be given to the drain line to ensure that the trap is properly seated when not in the bleed or vent position.

TROUBLESHOOTING

A common failure mode of steam traps is failure of the sealing device (i.e., plunger, disk, or valve) to return to a leak-tight seat when in its normal operating mode. Leakage during normal operation may lead to abnormal operating costs or degradation of the process system. A single $\frac{3}{4}$-in. steam trap that fails to seat properly can increase operating costs by \$40,000 to \$50,000 per year. Traps that fail to seat properly or are constantly in an unload position should be repaired or replaced as quickly as possible. Regular inspection and adjustment programs should be included in the Standard Operating Procedures (SOPs).

Most of the failure modes that affect steam traps can be attributed to variations in operating parameters or improper maintenance. Table 18.1 lists the more common causes of steam trap failures.

Operation outside the trap's design envelope results in loss of efficiency and may result in premature failure. In many cases, changes in the condensate load, steam pressure or temperature, and other related parameters are the root causes of poor performance or reliability problems. Careful attention should be given to the actual versus design system parameters. Such deviations are often the root causes of problems under investigation.

Poor maintenance practices or the lack of a regular inspection program may be the primary source of steam trap problems. It is important for steam traps to be routinely inspected and repaired to ensure proper operation.

19

PERFORMANCE MEASUREMENT AND MANAGEMENT

The measurement and subsequent management of organizational performance is necessary to determine whether goals and objectives are being met. Numerous books have been written detailing performance management, benchmarking, competitive analysis, or a multitude of other names. The purpose of this section is to provide an introductory overview of performance measurement and management. The first section will focus on measurement, and the next section will focus on management practices.

PURPOSE

The purpose of measuring performance is to help predict future action and performance based on historical data. Measuring performance helps identify areas that need management attention. On the other hand, measuring performance also highlights successful areas and accomplishments. Both are necessary to get a picture of how the organization is performing. Knowing the direction or trend in which the organization is headed is the first step in setting or correcting performance.

Developing raw data into useful information requires a skill set. Measurements must be reviewed on a regular basis to provide insight into the organization. Use of measurements can vary from one organization to another. The following pages provide an overview of using performance measurement.

BEST PRACTICES

To properly position the use of performance measurement, some general philosophical guidelines need to be developed. The following is intended to correctly position how performance measurement and management should be viewed.

- You can't measure everything.
- Performance management is like a gauge to equipment; it tells operating condition.
- Manage what you measure.
- Turn data into information, then into action.
- Indicators must tie into a "strategic business plan" and have purpose. The best organizations have most, if not all, of the following essential practices in place: (1) performance measurement/management system developed; (2) Key Performance Indicators (KPIs) driven by business initiatives and linked to strategic direction; (3) focus on internal trending measured and continuous improvement managed.
- Downplay only external comparison. External benchmarking data collected, analyzed, and "shared." "Core" critical success factors identified and measured/managed. "At a glance" performance designed systems and procedures.
- For both equipment and manufacturing processes.
- Posted and visible "on the fly." Accessible, user-friendly format, responsive to requests.
- Regular performance information review and updates, at least quarterly.
- Multiple recognition and reward systems designed and functioning. Simple to "digest" performance information.
- Must be in chart or graph form.
- With explanation and reveals trending measurement. What is "measured" gets "managed."

MANAGEMENT REPORTS

Every maintenance organization should have a reporting system, regardless of size or whether or not a computerized program supports the organization. Feedback from these systems provides assistance to the organization in determining whether its goals or objectives are being met, if it is satisfying customer needs, and if it is operating efficiently and economically. Last but not least, feedback helps to identify design and quality improvements. The following are examples of reports.

Broad Indicators

This group includes the ratio of maintenance costs to sales ratio of maintenance costs to value of assets maintenance expenditures by cost centers.

Work Load Indicators

This group includes current backlog, total backlog ratio of preventive maintenance to total maintenance, ratio of daily maintenance to total maintenance, ratio of work performed under blanket work orders (or charge numbers) to total maintenance, ratio of capital work to total maintenance, ratio of shutdown work to total maintenance, ratio of area maintenance to total maintenance, and ratio of craft equipment backlog.

Planning Indicators

This group includes jobs completed versus jobs planned, jobs completed versus jobs scheduled, estimated versus actual (effort-hours and cost) ratio of planned to unplanned jobs, ratio of emergency jobs to total jobs, and ratio of downtime to available "run-time."

Productivity Indicators

This group includes percentage of wasted time, maintenance labor costs versus maintenance material cost, and maintenance cost per unit of production.

Cost Indicators

This group includes work class or type percentages, actual maintenance cost as compared with budgeted costs, and percentage of maintenance administration cost to total maintenance cost.

These reports will be developed from information provided by all maintenance personnel. This is why it is so important to develop an efficient feedback system. The work order system is probably the most important feedback system in maintenance.

Equipment Efficiency

For a process industry, we recommend the following elements to be included in the analysis of equipment efficiency:

1. The percent of availability or uptime
   % AVAILABILITY OR TIME UP
   TIME UP × 100 A
   TIME UP + TIME DOWN
   TIME UP + TIME DOWN = 365 DAYS A YEAR

2. % PRIME QUALITY VOLUME
   PRIME QUALITY VOLUME × 100 Q
   PRIME QUALITY VOLUME + REJECTS

3. % OPERATING RATE OR SPEED
   TARGET CAPACITY × 100 S
   ACTUAL CAPACITY + SPEED LOSSES

OVERALL EQUIPMENT EFFICIENCY (OEE) IS

$$OEE = \%A \times \%Q \times \%S$$

Most plants or facilities use only A and Q, but it is important to also include speed losses in the analysis of overall equipment efficiency.

VALUE OF LOSSES/IMPROVEMENTS

Depending on the product you are manufacturing, the value of a 1% increase in OEE in a 400,000-ton-per-year plant corresponds to an increase in revenue of $1.6 million to $5.5 million per year. This corresponds to an increase in operating profits or contributions to cover fixed costs of $800,000 to $2.8 million per 1% increase in OEE, or an average per-minute capacity of 400,000 tons per year of $11.2 million to $39.2 million per Chapter 14.

The investment to achieve these savings is relatively small. We believe that most improvements can be accomplished by doing better with what you already have, and the key phrases are *planning of operations and maintenance* and implementation of *continuous improvement processes*.

Improvement efforts in maintenance performance alone can often affect more than half of the improvement potential, and increased integration between operations and maintenance improvement efforts will give you the full effect of your improvement efforts.

Maintenance efforts to increase OEE will almost always result in savings in maintenance costs in the range of 5–40%. Our experience data show that investments required to improve maintenance performance are in the range of 0.5–5% of the maintenance budget during the duration of an improvement project.

As a whole, maintenance improvement projects have a potential to pay back 5–15 times investments annually.

PRODUCTIVITY INDICATORS

During the past 10 years we have analyzed performance and developed productivity improvement plans for approximately 140 different paper machines and a large number of pulp mills including wood yards, bleacheries, recovery boilers, and power boilers, mainly in the United States, Canada, and Scandinavia. These analyses have been an in-depth focus on maintenance procedures and their impact on equipment efficiency and the productivity of mills.

In addition to these analyses of pulp and paper mills, improvement plans have been developed and implemented in more than 200 companies before 1985. To summarize, the following findings were based on a vast database and much experience.

Do not expect to learn anything new from this; with some additions, the bottom line is that planned maintenance is a key success factor and that planned maintenance cannot be achieved unless you have condition monitoring practices implemented to feed your planning procedures, which we refer to as *condition-based maintenance* (CBM).

The PQV/M Factor indicates how much Prime Quality Volume is being produced per $1,000 invested in maintenance. This is a Results Oriented Maintenance productivity indicator. The PQV/M Factor is the inverted value of maintenance cost per ton.

Equipment Efficiency is based only on % Uptime × % Rejects and does not include speed losses.

Poor performers only plan and schedule 10% of their work. An increase in planned and scheduled work improves equipment efficiency from a poor performance of 76% to 96%, an improvement of 20%. Concurrent with this, the PQV/M Factor increases from 18 to 44 Prime Quality Volume produced per $1,000 invested in maintenance. In summary, planned maintenance increases productivity and decreases costs for maintenance.

Concurrent with the increase in planned maintenance and OEE, the maintenance cost is gradually decreasing by 33%.

Other findings are that planned maintenance increases the technical life of the equipment. At 10% planned maintenance, the average life of electric motors is in the range of 3 to 7 years; pumps are 1 to 5 years, and bearings are 5 to 9 years. Excellent performers have two to three times longer technical life of their equipment.

Common for excellent performers is that they all were supported by well-organized professional maintenance resources.

Most of the investment is intended to help people to do better with what they already have.

MAINTENANCE OBJECTIVE

The prime goal of a maintenance operation is to provide equipment efficiency. The secondary goal is to deliver equipment efficiency as cost-effectively as possible. Unfortunately, it is common to see that most mills have turned this goal upside down and thus focus too much attention on cutting the maintenance cost. We all know that we can easily do this for a short period of time, but we have to pay back later.

The measurement goal for a maintenance organization should be:

PQV/M FACTOR long term maintenance effectiveness measure

$$\text{Prime Quality Volume} - X \quad \$1000 = PQV/M$$

Maintenance Cost
Maintenance Efficiency
Percent Unplanned   Percent Waiting Time U/W Factor
Maintenance Jobs   Related to Unplanned Jobs

The U/W Factor is suggested to be used as the day-to-day measurement of maintenance effectiveness. The factor should be recorded every day and followed up and compared with targets each week. An unplanned maintenance job is defined as a job that has to be started the same day it is initiated. An unplanned job always includes waiting time.

A brief explanation of the U/W factor follows.

Planned Work—Work orders written ahead of time for which someone has:

> 1. Estimated the necessary steps, skills, and manpower required to do the job

2. For each step, identified all stock materials (by stock number) and non-stock materials required, and ordered the non-stock items
3. Tracked material availability and analyzed personnel needs (by skill levels) and availability, as well as opportunity and availability of production equipment and special tools
4. Scheduled the job (at least one day ahead of time) based upon #3 and coordinated with production plans
5. Executed the job, without interruption, when scheduled

Unplanned Work—Work required to be done the same day as ordered and which might interrupt the daily schedule of work as outlined in "unplanned work."

Waiting (or Wasted) Time—Nonproductive time associated with unplanned work including:

1. Finding the personnel to do the work
2. The time it takes for the personnel to stop the job they are on and go to the unplanned job
3. Diagnosing the failure
4. Gathering the correct materials and tools to perform the job
5. Waiting on materials that are not on site
6. Coordinating people to work overtime (if necessary)

MAINTENANCE INPUT DATA REQUIRED TO MEASURE MAINTENANCE IMPROVEMENT EFFECTIVENESS

1. Percent Emergency Effort-hours (Period)
   Effort-hours spent on emergency work orders (period) × 100
   Total direct maintenance hours (period)

2. Percent Emergency and AD Unscheduled Effort hours (period)
   Effort hours spent on unscheduled work orders (period) × 100 = %
   Total direct maintenance effort hours (period)

3. Breakdown Equipment Time (period)
   Percent downtime caused by breakdown (period) × 100 = N
   Total downtime (period)

4. Percent Breakdown Repair Hours (period)
   Total effort hours on breakdown repairs (period) × 100 = %
   Total maintenance effort-hours (direct) (period)

5. Dollar Value of Breakdown Repairs (period)
   Cost breakdown repairs × 100 = !/0
   Total direct cost of maintenance

6. Percent Equipment Availability (period)
   Equipment run time × 100 = !/0
   Equipment run time and downtime

7. Percent Breakdowns Caused by Low Quality
   Breakdowns caused by low quality maintenance × 100 = !/0
   Total number of breakdowns

8. Dollar Value of Low Quality Maintenance Breakdowns
   Direct main cost and lost production cost × 100 %
   Total number of breakdowns

9. Maintenance Dollar Percentage Mill Book Investment
   Total maintenance cost × 100 = %
   Plant investment book

10. Percent Labor Costs to Material Costs
    Total maintenance labor cost × 100 =
    Total maintenance material cost

11. Percent Clerical Manpower Costs of Total Maintenance Cost
    Total clerical cost × 100 = %
    Total maintenance cost

12. Percent Supervision Cost of Total Maintenance Costs
    Total cost of supervision × 100 =
    Total maintenance cost

13. Maintenance Cost Percentage of Total Manufacturing Cost
    Cost of maintenance × 100
    Manufacturing cost

14. Maintenance Cost Percentage of Sales
    Total maintenance cost × 100
    Dollar value of sales

15. Cost of Maintenance Hour
    Total cost of maintenance × 100 = %
    Total effort hours worked

16. Breakdown Cost Component (period)
$$\frac{\text{Total breakdown cost} \times 100}{\text{Total production cost}} = \%$$

17. Percent Efficiency
$$\frac{\text{Total effort hours estimated for jobs} \times 100}{\text{Total effort hours spent on same job}} = \%$$

18. Percent Overtime
$$\frac{\text{Total overtime hours worked} \times 100}{\text{Total hours worked}} .0/0$$

19. Percent Work Orders Planned and Scheduled Daily
$$\frac{\text{Work order planned and scheduled} \times 100}{\text{Total work orders executed}} = \%$$

20. Percent Scheduled Hours vs. Hours Worked as Scheduled
$$\frac{\text{Hours worked as scheduled} \times 100}{\text{Total hours scheduled}} \%$$

21. Percent Scheduled Hours vs. Total Hours Worked
$$\frac{\text{Hours scheduled} \times 100}{\text{Total hours worked}} \%$$

22. Percent Work Order Executed As Scheduled
$$\frac{\text{Work orders executed as scheduled} \times 100}{\text{Total work orders scheduled}} = \%$$

23. Compliance with Estimated Cost
$$\frac{\text{Jobs executed at or within 15\% of estimated cost} \times 100}{\text{Total estimated jobs executed}} = \%$$

24. Ratio Maintenance Planners
$$\frac{\text{Total hourly personnel} \times 100}{\text{Total maintenance planners}} = \%$$

25. Percent Estimated Coverage
$$\frac{\text{Number of planned work orders} \times 100}{\text{Number of work orders completed}} \%$$

26. Percent Activity Level of Maintenance Craftsmen
$$\frac{\text{Direct time working on work orders} \times 100}{\text{Work orders released for work}} \%$$

27. Work Order Turnover Percentage
Number of work orders completed during period × 100 %
Number of work orders awaiting release

28. Current Crew Backlog (Weeks)
Effort-hours ready to release = weeks
One crew week (effort hours)

29. Total Backlog by Crew (Weeks)
Total effort-hours ready to work = weeks
One crew week expressed in effort hours

30. P.M. Coverage Percentage
    A. Effort hours spent on P.M. work orders (period)
    B. Total effort hours worked (period)
    C. P.M. inspections incomplete (period)
    D. P.M. inspections scheduled (period)
    E. Repair jobs resulting from inspection (period)
    F. P.M. inspections completed

WHERE TO START

No two organizations have the exact same performance indicators. This is because situations are different. Most organizations attempt to or think that they need to track 50 or more indicators. This is a waste of time and effort. The focus should be on quality, not quantity.

To assist organizations in this transformation process, EDCON Inc. has prepared a recommended list of performance management indicators. These indicators should form "the core" of the performance measurement and management system.

This core group is supported by other indicators that are tracked in support of the core group. The supporting group might be tracked for several months, then replaced by another indicator. However, the core group of indicators always (or almost always) remains the same. The following is a recommended list of core group indicators.

Competitive indicators focus more on industry and "macro" internal measurement. Productive measures focus on maintenance and quality production. Operational indicators focus more on equipment. The combination of all three indicators will provide an excellent "snapshot" of organizational performance.

- Competitive
  - Sales / revenue / production (prime quality) volume comparison
  - Prime quality production output per manufacturing expenses
    -Measured in output (i.e., tons, widgets, lbs, bbl.)
  - Prime quality production output per $ 1,000 invested in mainten-ance

- Productive
  - Prime quality production output per total labor hours worked
    -Can divide labor hours out by maintenance-production-total
  - Maintenance budget management
  - Work order priority distribution percentages
  - Backlog aging (30−60−90−90+) "maintenance accounts payable"
  - Backlog hours by area and craft

- Operational
  - Availability (uptime) × Production (operating rate or speed)
    × Quality = OEE—Overall Equipment Effectiveness
    "Big 5" equipment management analysis
    "Equipment of the month" focus

PERFORMANCE MANAGEMENT

The "second key" of developing an effective performance measurement system is performance management. A best practice approach is "to manage what you measure."

The following is an overview of the management aspects of performance management. Several key performance measures will be reviewed and the management implications discussed.

Backlog

Backlog is a performance indicator that gives insight into the ability of the maintenance organization to manage and control work. It can also be used, with other indicators, to gain insight into staffing needs, training requirements, and outside contractor assistance.

Backlog level should be managed and controlled around a range of 2 to 8 weeks of work per craftsperson. This can also be tracked in hours per craftsperson. Backlog is designed to "feed" planned work into the scheduling machine.

If backlog levels are too low, efficient planning, parts ordering and inventory management, and scheduling of the work force will never approach world-class levels. On the other hand, if backlog levels remain low and the emergency work level is low (i.e., 2%), the planning function might need to scope and plan more jobs or the work force could be too large or overstaffed.

If the backlog exceeds the 6- or 8-week level, high-priority work may not be performed on time. This might be due, in part, to poor scheduling, low response level by maintenance, lack of an objective work order priority system, or too little staff. Temporary increases in overtime or the use of outside contractors may be a short-term fix. If backlog levels trend consistently high, an increase in staffing levels may be necessary.

A "best practice" recommendation is to have backlog listed by area and craft. Examples are attached.

Preventive maintenance, project, and construction backlog should be tracked on a separate chart. Backlog might be tracked by "total" backlog that would include the above classifications. It is recommended that another set of reports be developed that would track the backlog by each individual classification. This will provide greater insight into the operations of the organization.

Backlog Age

Backlog age (30−60−90+ day report) is a measurement of the number of work orders in the backlog sorted out by the age or length of time these jobs have "waited" until they were completed.

Backlog age is best characterized to show the raw or actual number of work orders. The following example shows this method.

PRIORITY

The goal should be that no critical or routine plannable jobs should be on the backlog for over 90 days. These jobs have already gone through the entire work order process of write-up, approval, planning, parts ordered, and awaiting scheduling. This is a joint effort between operations and maintenance to manage their work. The reader is encouraged to refer back to the backlog management and operational asset management chapters in this text for additional information.

Schedule Compliance

Schedule compliance uses the daily and weekly schedule that was developed in the scheduling chapter and in the operational assets management chapter. Schedule compliance simply quantifies the question "Did we accomplish what we wanted to?" World-class schedule compliance is 95%. Ideally, schedule compliance would be 100%. In a real application, where "things" happen, schedule compliance of 90% is certainly a realistic goal. If scheduled work is constantly bumped (schedule creep) by emergency work or "emotional priorities," then the maintenance and operations daily/weekly schedule is just a "paper exercise." The reader is encouraged to review the sections on daily/weekly schedule use and operational asset management.

Analyzing Trends

After reaching a planned/scheduled execution of over 90% of all maintenance hours, work on more long-term strategies by analyzing the following trends:

- Backlog hours
- Overtime hours
- Contractors' hours

Assuming that a maintenance organization is planning, scheduling, and executing not less than 90% of all maintenance accordingly and backlog, overtime, and contractors' hours are all up, then you need to add people.

If your planning and scheduling level is more than 90% and all of the above trends go down, your strategy should not be to reduce the number of people but to switch your focus to continuous improvement. This strategy is important; if you do not clearly document it, convey it to your organization, and show long-term commitment, you will not gain support from your people. Instead, you will notice resistance or a lack of support. These ideas have been mentioned previously.

Work Order Priority Analysis

Work order priority analysis measures the distribution of work entering the work management system based on priority. This measure indicates several things, but one of the more valuable indications of the overall "health" of the maintenance management system is, for example, the goals should be fewer emergency and critical work orders and more routine or plannable and scheduling jobs.

Stores and Material Management

As previously mentioned, stores plays a vitally important role in the overall maintenance management process. Several key indicators of the most important stores-related indicators are:

- Inventory accuracy
- Stock-outs (active inventory)
- Service level (active inventory)
- Hot shot/expedite
  - Percent of all deliveries
  - Dollar value
- Inventory issue frequency

For example, stock-out percentage is the fraction of occurrences in which the storeroom cannot fill a requisition list for parts or materials because it is out of stock.

Stock-outs should fall in the range of 0–2%.

Another recommended stores performance indicator is turnover. Turnover is the calculation of the average throughout the storeroom (expressed in months).

$$\text{Turnover} = \frac{\text{Total Value of the Storeroom}}{\text{Total Value Withdrawn in a Year}} \times 12 \text{ months/year}$$

If turnover is high, it is an indication that the stores value is inflated. If it is low, excessive stock-outs may be the result. The formula is best for broad categories of the stores inventory, as different categories will reflect different turnover rates. Common categories for storeroom inventory are:

- Recommended
- Turnover
- Category

Insurance items, one-of-a-kind spare equipment; large assemblies; spare parts, for pumps, gear boxes, etc.; hardware, such as nuts and bolts; supplies, housekeeping materials, etc. Little or no turnover every 2 years, every year, every 6 months, every 1 to 3 months.

The turnover rate should fall within the range of 6–12 months.

In addition to tracking inventory by turnover, another effective indicator is tracking inventory by issue frequency. This indicator will highlight inventory based on frequency of use or issue. The goal is to increase the amount of issue inventory and decrease the levels of slow, stale, and no activity. Typical performance indicator ranges are listed below.

- Classify inventory by issue frequency
- Active usage last 12 months   70%
- Slow, no usage last 13 months   15%
- Stale, no usage last 13–24 months   10%
- No activity no usage last > 25 months   5%

It is recommended that action be taken to address why inventory is "on the shelf" yet hasn't been issued.

COMPOSITION OF INVENTORY CARRYING COSTS

- Capital, opportunity or borrowing costs   10–15%
- Cost of operating warehouse space, property tax
- Energy cost, insurance
- Cost of space occupancy, rent, and depreciation
- Cost of inventory shrinkage, obsolescence, damage, theft, and contamination   5–10%
- Inventory tax   1–2%
- Management costs, labor, computer support, and administration 5–10%
- Total of inventory carrying costs per year   30–40%

Shutdown and Project Management Guidelines

Suggested performance indicators include:

    a. Schedule performance
      On time
    b. Budget performance
      +/− 5% of allocated shutdown budget
    c. Equipment "out of service" downtime
      Reduced 5% to 25% versus previous shutdown projects
      Big impact on pre-scheduling and performing smaller activities throughout the year

Performance Management Reporting

It is highly recommended that the maintenance organization develop a weekly report. This report must be reviewed on a regular basis. Weekly is suggested, and a "best practice" is to have an open-door review policy.

The other suggested forum for review of maintenance management performance information is at the weekly operational asset management meeting.

This is an ideal forum for integrating performance management information into the operational fabric of the organization.

The examples provided illustrate how different organizations developed different unique methods for reporting organizational performance indicators.

GLOSSARY

Actuator system

The system that provides a reliable means to engage and disengage the mating parts of a clutch that provide power transmission.

Addendum

Distance the tooth projects above, or outside, the pitch line or circle.

Additive

Chemical added in small quantities to a petroleum product to enhance, as opposed to alter, certain desirable properties.

Alignment

The process of adjusting a piece of machinery both vertically and horizontally so that its shaft has a common axis of rotation at operating equilibrium with the shaft of the machine to which it is coupled.

Ambient

Surrounding condition, especially pertaining to ambient temperature.

Amplitude

The magnitude or size of a quantity such as velocity, displacement, acceleration, etc., measured by a vibration analyzer, displacement probe, etc.

Angularity

The angle between two shaft centerlines, which generally is expressed as a slope (i.e., rise/run) rather than in degrees.

Arbor press	A machine used for forcing an arbor or a mandrel into drilled or bored parts preparatory to turning or grinding. Also known as a mandrel press.
Articulated structure	A structure in which relative motion is allowed to occur between parts, usually by means of a hinged or sliding joint or joints.
Axial	Of, on, around, or along an axis (straight line about which an object rotates) or center of rotation.
Axis	Straight line about which an object rotates.
Babbitt metal	Any of the white alloys composed primarily of tin or lead and of lesser amounts of antimony, copper, and perhaps other metals, and used for bearings.
Backlash	Gears are the play between teeth that prevents binding.
Balancing	Process of bringing an imbalanced device into equilibrium or balance by the addition or subtraction of known weight to counterbalance the forces caused by an unknown weight.
Balancing arbor	The shaft on a balancing machine on which the rotor to be balanced is mounted.
Bar stock	Common term for non-specific iron or steel material. Not to be used for keys, which require specific shearing properties. Also see "Key Stock."
Bearing	A machine element that supports a part, such as a shaft, that rotates, slides, or oscillates in or on it. Three major types are radial or journal, thrust, and guide bearings.
Bearing, conrad	Type of ball bearing that is classified as a single-row radial, non-filling slot bearing. Also known as a deep-groove bearing.
Bearing, deep-grooved	See Conrad bearing.

Bearing, fluid-film	Another name for plain bearings.
Bearing, guide	A bearing that guides a machine element in its lengthwise motion, usually without rotation of the element.
Bearing, radial	A bearing that supports loads acting radially and at right angles to the shaft center line. These loads may be visualized as radiating into or away from a center point like the spokes on a bicycle wheel.
Bearing, thrust	A bearing that prevents lengthwise or axial motion of a rotating shaft.
Bellows coupling	A material-flexing coupling that consists of two shaft hubs connected to a flexible bellows. The bellows allows for minor misalignment while operating under moderate rotational torque and shaft speeds.
Belt	A flexible band used to connect pulleys or to convey materials by transmitting motion and power.
Belt drive	A means of power transmission between shafts by means of a belt connecting pulleys on the shafts.
Bending tension	The force created in a belt due to bending (e.g, around a pulley).
Bevel gear	Is a cone.
Bore	The hole in a rotor through which the shaft is placed.
Bore	The hole in a rotor through which the shaft is placed.
Brinell number	A hardness rating obtained from the Brinell test; expressed in kilograms per square millimeter (kg/mm^2).
Brinell test	A test to determine the hardness of a material, in which a steel ball one centimeter in diameter is pressed into the material with a standard force (usually

3,000 kilograms). The spherical surface area of indentation is measured and divided into the load and the results are expressed as Brinell number.

Broaching fluid	Fluid used in the broaching process, which is used to restore the diameter of a bore hole by reaming with a multiple-tooth, bar-like cutting tool. The teeth are shaped to give a desired surface or contour, and cutting results from each tooth projecting farther than the preceding one.
Burr	Thin, ragged fin left on the edge of a piece of metal by a cutting or punching tool (as on a bore hole).
Burr	Thin, ragged fin left on the edge of a piece of metal by a cutting or punching tool (as on a bore hole).
Bushing	A removable piece of soft metal or graphite-filled sintered metal that lines a support for a shaft. The core of a compression rigid coupling is comprised of a slotted bushing that has been machine bored to fit both ends of the shafts.
Butt cut	Square cut.
Butt joint	A means of joining plates in the same plane by a cover plate or butt strap riveted to both plates by one or more rows of rivets.
Calibration	The process of determining, by measurement or comparison with a standard, the correct value of each scale reading on a meter or other device, or the correct value for each setting of a control knob.
Carbonaceous	Relating to or composed of carbon.
Catalyze	Modify the rate of a chemical reaction with a catalyst.
Centrifugal force	An outward pseudo-force in a reference frame that is rotating with respect to an

inertial reference frame. The pseudo-force is equal and opposite to the centripetal force that acts on a particle stationary in the rotating frame.

Centrifugal pump — A machine for moving a liquid by accelerating it radially outward by an impeller to a surrounding volute casting.

Centripetal force — The radial force required to keep a particle or object moving in a circular path, which can be shown to be directed toward the center of the circle.

Chain — A flexible series of metal links or rings fitted into one another.

Chain belt — A chain-based device used to convey objects or for transmitting power.

Chain coupling — A mechanical-flexing flexible coupling that provides a means of transmitting proportionally high torque at low speeds and compensates for minor shaft misalignment. It is composed of hubs having teeth and a connecting chain.

Chain drive — Flexible device for power transmission, hoisting, or conveying. Consists of an endless chain whose links mesh with toothed wheels fastened to the driving and driven shafts.

Chain pitch — Distance between the centerlines of two successive pins.

Chattering — A mode of operation of a relay-type control system in which the relay switches back and forth infinitely fast.

Chordal — Of or pertaining to a line connecting any two points across a circle.

Circular pitch — The distance form a point on one tooth to a corresponding point on the nest tooth, measured along the pitch line or circle.

Circumference	The length of a circle or other continuous path measured along the outside edge.
Circumferential	Relating to the circumference, the length of a circle.
Clearance	Amount by which the addendum of a gear tooth exceeds the addendum of a matching gear tooth.
Clutch	A coupling designed to transmit intermittent power to a driven unit. Types: positive (one way and two way) and friction.
Coil spring	A helical or spiral spring such as one of the helical springs used over the front wheels in an automotive suspension.
Colinear	Condition when the rotational centerlines of two mating shafts are parallel and intersect (i.e., join to form one line).
Compressed packing seal	See "Packed Stuffing Box."
Compression	A force having the effect of reducing volume or shortening length due to pressure.
Compression rigid coupling	A coupling that depends on friction for transmission of power, which is comprised of three pieces: a compressible core and two encompassing coupling halves that apply force to the core.
Compression-type coupling	See "Elastomeric Coupling." Compression-type couplings may be fitted with projecting pins, bolts, or lugs to connect the components.
Concentricity	When the smaller of two circular, cylindrical, or spherical shapes is centered within the larger one.
Converging fluid film	Fluid film sustained by motion.
Correction plane	Plane in which a balancing correction is made by adding or removing weight.

Couple	Connecting two axially oriented shafts of a driver and a driven unit using a coupling.
Couple imbalance	Imbalance caused by two equal non-colinear forces that oppose each other angularly (i.e., 180 degrees apart).
Coupling	Mechanical devices, classified as either rigid or flexible, used to connect two axially oriented shafts of a driver and a driven unit.
Crazing	A network of fine cracks on or under the surface of a material.
Critical speed	The angular speed at which a rotating shaft becomes dynamically unstable with large lateral amplitudes due to resonance with the natural frequencies of lateral vibration of the shaft.
Cross-section	An exposure or cut that exposes internal features.
Crowned	Having a point where the thickness or diameter increases from edge to center.
Crush	Bearings are slightly longer circumferentially than their mating housings, which upon installation is elastically deformed or "crushed" to ensure good back contact.
Datum dimensions	Standards specifying v-belt dimensions, such as length.
Dedendum	Depth of a tooth space below, or inside, the pitch line or circle.
Deflection	Elastic movement or sinking of a loaded structural member.
Deflection force	Force applied with a belt tension tester perpendicular to the center of the span. This force should be large enough to deflect the belt $\frac{1}{64}$-inch for every inch of span length.

Degree of freedom	Any one of the number of ways in which the space configuration of a mechanical system may change.
Demulsibility	The ability to demulsify or break liquid-liquid emulsions or to prevent them from forming.
Detergent	Synthetic cleansing agent resembling soap in its ability to emulsify oil and hold dirt, and containing surfactants that do not precipitate in hard water.
Diametrical pitch	A gear tooth design factor expressed as the ratio of the number of teeth to the diameter of the pitch circle measured in inches.
Diaphragm	A material-flexing coupling designed to provide torsional stiffness while allowing flexibility in axial movement.
Differential driving	The condition caused by badly worn pulley grooves causing belts in the worn groove to ride lower than other belts on the same pulley. The higher-riding belts travel faster than their lower-riding counterparts.
Dishing	Wear on the sides of grooves in a pulley, which results in a shallow concave surface.
Dispersant	An additive that can hold finely ground materials in suspension.
Double seal	Two seals on the same shaft, usually having a fluid pumped between them for cooling or pressure boundary purposes.
Drive shaft	The shaft that supplies power to a unit.
Drive sprocket	The drive sprocket is attached to the source of power (i.e., electric motor, crankshaft, etc.).
Drive train	Connects a motor to a propeller or driven axle. May include drive shaft, clutch, transmission, and differential gear. Also known as a power train.

Driven unit	A device (e.g., fan, pump) that is driven by a power source such as a motor, turbine, etc.
Driver	Power source such as a motor, turbine, etc.
Dynamic imbalance	Imbalance in two separate planes at an angle and magnitude relative to each other not necessarily that of pure static or pure couple.
Dynamics	That branch of mechanics that deals with the motion of a system under the influence of forces, especially those that originate outside the system under consideration.
Eccentric	The condition when a disk or wheel has its axis of revolution displaced from the center so that it is capable of imparting reciprocating motion.
Eccentricity	The distance of the geometric center of a revolving body from the axis of rotation.
Eccentricity	The distance of the geometric center of a revolving body from the axis of rotation or not having the same center. Eccentricity of the shaft with the stuffing box bore alters the hydraulic loading of the seal faces, reducing seal life and performance.
Elastic deformation	Reversible alteration of the form or dimensions of a solid body under stress or strain.
Elastomer	A polymeric material, such as a synthetic rubber or plastic, which at room temperature can be stretched under low stress to at least twice its original length and upon immediate release of the stress, will return with force to its approximate original length.
Elastomeric	A coupling that can compensate for minor misalignments because it consists of two hubs separated and connected by an elastomeric element, which can be placed in

either shear or compression. Used in light- or medium-duty applications running at moderate speeds.

End play	Lateral or axial shaft movement.
Entrainment	Entrapment of tiny air bubbles in a fluid.
Environmental controls	Auxiliary systems that supply cooling, heating, and/or lubrication to mechanical seals.
Equilibrium	Condition when no change occurs in the state of a system as long as its surroundings are unaltered.
Face readings	Dial-indicator readings obtained as the shafts are rotated and the dial-indicator stem is parallel to the shaft centerline with a point of contact on the (face of the coupling??).
Fatigue	Failure of a material by cracking resulting from repeated or cyclic stress.
Fatigue life	The number of applied repeated stress cycles a material can endure before failure.
Fillet	Concave transition surface between two otherwise intersecting surfaces. Also refers to a corner piece at the juncture of perpendicular surfaces to lessen the danger for cracks.
Flange	Projecting rim of a mechanical part.
Flanged rigid coupling	A rigid coupling comprised of two flange-halves, one located on the end of the driver shaft and the other on the end of the driven shaft. These halves fasten together with bolts, providing positive transmission of power.
Flexible coupling	A coupling that allows components to slide or move relative to each other. Although clearances permit movement within specified tolerance limits, flexible couplings are not designed to compensate for major

	misalignments. Types: mechanical- and material-flexing.
Flexible rotor	A long rotor running at a speed greater than 80 percent of the rotor's first critical speed.
Flexible shaft	A material-flexing coupling generally used on small equipment applications that do not come under high torque loads.
Floating-shaft coupling	A special application coupling that accommodates greater misalignment or the ends of the driver and driven shafts that have to be separated by a considerable distance. Also known as "Spacer Coupling."
Flywheel clutch	A one-way clutch that transmits torque in one direction and disengages in the opposite direction (e.g., bicycle freewheel).
Force imbalance	See "Static Imbalance."
Formulation	The particular mixture of base chemicals and additives required for a product.
Fretting	Wear that occurs when cyclic loading, such as vibration, causes two surfaces in intimate contact to undergo small oscillatory motions with respect to each other.
Friction	Resistance to sliding, a property of the interface between two solid bodies in contact. Friction wastefully consumes energy, and wear changes dimensions.
Friction clutch	A clutch that transmits torsional power from a driver to a driven unit and brings both shafts to the same speed. Relies on friction to transmit power so the transition from disengaged to engaged is more gradual than that of a positive clutch.
Gear	A form of disc, or wheel, that has teeth around its periphery for the purpose of providing a positive drive by meshing the

teeth with similar teeth on another gear or rack.

Gear coupling	A mechanical-flexing flexible coupling that transmits proportionally high torque at both high and low speeds. Torque is transmitted through gear teeth. Sliding action and ability for slight adjustments in position comes from a certain freedom of action provided between the two sets of teeth. Also known as "gear tooth coupling."
Gear lash	Amount by which a tooth space exceeds the thickness of the engaging tooth on the operating pitch circles. Also referred to as gear backlash.
Gearbox	The gearing system by which power is transmitted from the engine to the rotating shaft or axle.
Gland	A device for preventing leakage at a machine joint, as where a shaft emerges from a vessel containing a pressurized fluid. See "Packed Stuffing Box."
Grease	A solid or semi-solid lubricant comprised of up to 90 percent oil (with mineral oil being the most common lubricating fluid used), a thickening or gelling agent such as soap, and other ingredients such as additives and dyes.
Grout	Fluid mixture of cement and water, or a mixture of cement, sand, and water.
Growth factor	Coefficient of thermal expansion, mils/inch/°F
Half key	Key having full key length, but only half key depth.
Halogenated	Containing halogens in the chemical formula. Halogens are any of the halogen family, consisting of fluorine, chlorine, bromine, iodine, and astatine.

Helical	Pertaining to a cylindrical spiral, for example a screw thread.
Helical gears	Gear wheels running on parallel axes, with teeth twisted oblique to the gear axis.
Helix angle	The angle at which the gear teeth are cut.
Herringbone gear	The equivalent of two helical gears of opposite hand placed side by side.
Hold-down bolts	The bolts and nuts that are used to secure each foot of a machine its base.
Horsepower	The unit of power in the British engineering system, equal to 550 foot-pounds per second, approximately 745.7 watts. Abbreviated hp.
Hub	(1) The cylindrical central part of a wheel, propeller, or fan that fits over a shaft. (2) A piece in a lock that is turned by the knob spindle causing the bolt to move.
Hydraulic	A coupling designed to provide a soft start, with gradual acceleration and limited maximum torque for fixed operating speeds. Used in applications such as compressors that undergo torsional shock from sudden changes in equipment loads. Also referred to as "Fluid Couplings."
Hydraulic forces	Forces that result from the action of fluids.
Hydraulic jack	A jack in which force is applied by fluid pressure.
Hydraulic lift	An elevator or elevator-like device operated by fluid pressure.
Hydrodynamic instability	Instability in the hydrodynamic lubrication state, which occurs when the pressures developed in a converging fluid film (i.e., a film sustained by motion) are sufficient to support the bearing load.
Hydrolytic stability	The ability to withstand the hydrolytic reaction wherein water effects a double decomposition with another compound,

hydrogen going to one compound and hydroxyl to another.

Idler pulley	Used to guide and tighten the belt or chain of a system.
Imbalance	Condition when there is more weight on one side of a center line than the other.
Imbalance	Condition when there is more weight on one side of a center line than the other.
Impeller	The rotating member of a turbine, blower, fan, axial or centrifugal pump, or mixing apparatus. Also known as a rotor.
Impregnated	The condition whereby a liquid substance (such as oil) has been forced into the spaces of a porous solid (such as a metal) in order to change its properties.
Indicator sag	The bending of the dial indicator mounting hardware that occurs when rotated from the top to the bottom position.
Inertia	The property of an object by which it remains at rest or in uniform motion in the same straight line unless acted upon by some external force.
In-place balancing	Process of balancing a rotor without taking it out of the machine.
Interference fit	A fit wherein one of the mating parts of an assembly is forced into a space provided by the other part.
Intermediate drive	A belt or chain drive system interposed between a driver and a driven unit.
Jack bolts	Horizontally positioned bolts on the machine base located at each foot of the machine used to adjust the horizontal position of the machine.
Journal	That part of a shaft, axle, spindle, etc., that is supported by and turns in a bearing.
Journal box	A metal housing for a journal bearing.

Kevlar	Trade name for an aromatic polyamide fiber of extremely high tensile strength and greater resistance of elongation than steel.
Key	A piece of material, usually metal, placed in slots (keyway) cut into two axially oriented parts to mechanically lock them together.
Key stock	The material used to make keys. Properly sized key stock must be used with all keyways. Key stock materials include AISI 1018 and AISI 1045.
Keyed-shaft rotor	Any rotating element whose shaft incorporates a keyed connection in coupling two axially oriented shafts of a driver and a driven unit.
Keyseat	Axially located slot in a shaft into which a key is fitted in order to make a keyed coupling connection.
Keyway	A slot machined in a hub that is used in making a keyed coupling connection between two axially oriented components.
Kinetic energy	Energy associated with motion.
Laminar	Viscous streamline flow without turbulence.
Laminated disk-coupling	A material-flexing coupling that consists of hubs connected to a single or series of flexible disks that allow movement. Reduces heat and axial vibration from transmitting between the driver and driven unit.
Lantern ring	A grooved, bobbin-like spool piece that is situated exactly on the center-line of the seal water inlet connection to the gland. (A ring or sleeve around a rotating shaft; an opening in the ring provides for forced feeding of oil or grease to bearing surfaces; particularly effective for pumps handling liquids.)

Laser	An active electron device that converts input power into a very narrow, intense beam of coherent visible or infrared light, which radiates in phase.
Lateral	Of, at, on, or toward the side
Lead	Term referring to keyway manufacturing tolerance between the center lines of the key seat and the shaft.
Lubricant	A lubricant is a gas, liquid, or solid used to prevent contact of parts in relative motion, reducing friction and wear. Lubricants also provide machine cooling, rust prevention, prevention of solid deposits on close-fitting parts, and power transmission.
Lubrication	The use of a lubricant to maintain a fluid film between solid surfaces to prevent their physical contact.
Lubrication, fluid-film	Full-fluid film lubrication is the ideal state where the film remains thick and prevents contact between the surface peaks, which are apparent in a microscopic examination of two surfaces
Lubricity	The ability of a material to lubricate.
Machine train	A series of machines containing both driver and driven components.
Magnaflux	A material examination technique using magnetic fields, which detect holes and cracks.
Material-flexing coupling	A coupling that provides flexibility by permitting certain components to flex. Use governed by the operational fatigue limits of the materials (e.g., metal, plastics, or rubber) used to make the flexing elements.
Mechanical imbalance	See "Imbalance."
Mechanical seal	A device that incorporates a coil spring sitting against the back of the pump's impeller and pushing the packing "O" ring

against the seal ring. It is used on centrifugal pumps or other type of fluid handling equipment in applications where shaft sealing is critical. Requires no manual adjustment to maintain seal.

Mechanical-coupling

A coupling that allows flexibility in the components by permitting them to move or slide relative to each other. Types: chain, gear, and metallic grid.

Metallic-grid coupling

A combination of mechanical-flexing and material-flexing type couplings. Also referred to as "Combination Coupling." Compact coupling unit capable of transmitting high torque at moderate speeds. The flexing of a grid within grooved slots provides torsional resilience.

Micrometer

A unit of length equal to one-millionth of a meter.

Microprocessing unit

A microprocessor with its external memory, input/output interface devices, and buffer, clock, and driver units. A microprocessor is a single silicon chip on which the arithmetic and logic functions of a computer are placed.

Mil

One one-thousandths of an inch (1 mil = 0.001 inch).

Millwright

A person who plans, builds, or sets up the machinery for a mill or a person who repairs milling machines.

Misalignment

When two parts (e.g., shafts) to be coupled do not lie in the same axial plane.

Modulus of elasticity

Ratio of the stress to the strain, such as Young's modulus, the bulk modulus, or the shear modulus. Also known as coefficient of elasticity, elasticity modulus, and elastic modulus.

Mole

An amount of substance of a system that contains as many elementary units as there are atoms of carbon in 0.012 kilogram of the pure nuclide carbon-12. The elementary unit must be specified and may be an atom, molecule, ion, electron, photon, or even a specified group of such units.

Molecular weight	The sum of the atomic weights of all the atoms in a molecule. Units: grams per gram-mole, kilograms per kilogram-mole. Also known as relative molecular mass.
Nominal	Not real or actual, theoretical.
Offset	Term referring to keyway manufacturing tolerance between the center lines of the keyed components.
One-way clutch	A type of positive clutch. See "Flywheel Clutch."
Operating condition	Defines the physical requirements, dimensions, and type of coupling needed in a specific application. Envelope information includes: shaft sizes, orientation of shafts, required horsepower, full-range of operating torque, speed ramp rates, and any other data that directly or indirectly affects the coupling.
Orifice	An aperture or hole.
Oscillate	To move back and forth with a steady uninterrupted rhythm.
Overhung load	When there is no room to support a piece at both ends, an overhung load results.
Packed box	See "Packed Stuffing Box."
Packed stuffing box	A packed, pressure-tight joint for a shaft that moves through a hole to reduce or eliminates fluid leakage. The joint is a box containing a soft pliable material or packing that is compressed into rings encircling the drive shaft. Requires periodic maintenance to maintain compression. Care must be taken to

	prevent shaft wear. Also referred to as a "packed box" or "stuffing box."
Phase shift	The difference between the angle of deflection of a rotating part at two different measurement positions on the part or at the same measurement position at different times.
Pillow block	Type of premounted bearing.
Pinion	The smaller of a pair of gear wheels or the smallest wheel of a gear train.
Pitch	The distance between similar elements arranged in a pattern or between tow points of a mechanical part, as the distance between the peaks of two successive grooves on a disk recording or on a screw.
Pitch length	The length of the neutral axis of the belt.
Pitting	Selective localized formation of rounded cavities in a metal surface due to corrosion or sinking a pit.
Plane	A surface such that a straight line that joins any two of its points lies entirely in that surface. A place is defined by three points.
Polar plot	Circular diagram that represents the orbit of a shaft or rotor element around its centerline.
Positive clutch	Hubs with interlocking teeth that transmit positive (i.e., no slippage) power from the driver to the driven machine component. Types: One-way and two-way.
Positive power	Power generated by a mechanism that guarantees absolutely no slippage. Such systems have a direct mechanical link between the drive and the driven shaft (e.g., gears that incorporate interlocking teeth).
Precision balancing	Balancing to tolerances that produce velocities of 0.01 ins./sec. (0.3 mm/sec.) and lower.

Pressure angle	The sides of each tooth incline toward the center top at an angle.
Pulley	A wheel with a flat, round, or grooved rim that rotates on a shaft and carries a flat belt, V-belt, rope, or chain to transmit motion and energy.
Pyrometer	Any of a broad class of temperature-measuring devices, including radiation pyrometers, thermocouples, resistance pyrometers, and thermistors.
Rabbet fit	A non-supporting, mechanical positioning device located between the motor flange and the baseplate designed to automatically center the motor during installation.
Race	Either of the concentric pair of steel rings of a ball or roller bearing.
Raceway	A channel used to hold a bearing race or one of the concentric steel rings of a ball or roller bearing.
Rack	A bar containing teeth on one face for meshing with a gear.
Radial	Extending from a point or center in the manner of rays (as the spokes of a wheel are radial).
Reciprocation	The action of moving back and forth alternately.
Residual imbalance	Imbalance of any kind that remains after balancing.
Resonance	A large amplitude vibration in a mechanical system caused by a small periodic stimulus of the same or nearly the same period as the natural vibration period of the system. Higher levels of input energy can result in catastrophic, near instantaneous failure of the machine or structure.

Rigid coupling	A coupling that permits neither axial nor radial relative motion between the driver and driven unit shafts. Types of rigid couplings are flanged, split, and compression.
Rim readings	Dial-indicator readings obtained as the shafts are rotated and the dial indicator stem contacts the shaft at a 90-degree angle.
Ring	A tie member or chain link; tension or compression applied through the center of the ring produces bending moment, shear, and normal force on radial sections.
Rise	Represents shaft offset when right triangle concepts are used to determine misalignment corrections. "Run" represents the true, or target, shaft centerline. The angle or degree of deviation of the shaft from the target centerline is determined by the inverse tangent of the slope, which is rise divided by run.
Rivet	Permanent joining of two or more machine parts or structural members by means of rivets, short rods with a head on one end. The rod is inserted through aligned holes in parts to be joined, and the protruding end is pressed or hammered to form a second head.
Rotor	Any rotating part.
Run	See "Rise."
Run-in	A period of 24 to 48 hours required to accommodate stretching and settling of a new belt.
Runout	Axial or radial looseness. A measure of shaft wobble caused by being off center.
Sacrificial sleeve	A sleeve over the drive shaft designed to mechanically wear, protecting the drive shaft, and be easily replaced.

Sag	Measured by pulling the bottom span of a chain taut, allowing all of the excess chain to accumulate in the upper span, and placing a straightedge on top of the sprockets. Chain sag should be approximately $\frac{1}{4}$ inch for every 10 inches of sprocket centers.
Sag factor	A correction that must be made to dial indicator readings. See "Indicator Sag."
Saponification	The process of reacting a fatty oil with alkalies (e.g., aluminum, calcium, sodium, or lithium hydroxide) to make soap.
Seal	A means of preventing migration of fluids, gases, or particles across a joint or opening.
Shaft	A cylindrical material element, usually metal and solid, which rotates and transmits power.
Shear	A straining action wherein applied forces produce a sliding or skewing type of deformation. Force acts parallel to a plane as distinguished from tensile or compressive forces, which act normal to a plane.
Shearing	Slipping or sliding of one layer of a substance relative to its adjacent layer.
Shear-type coupling	A type of elastomeric coupling in which the elastomeric element may be clamped or bonded in place, or fitted securely to the fitted sections of the hubs. See "Elastomeric Coupling."
Shim	Thin piece of material placed between two surfaces to obtain a proper fit, adjustment, or alignment.
Sludge	Any semi-solid waste from a chemical process.

Soft foot	Condition when all four of a machine's feet do not support the weight of the machine
Soft foot	Condition that exists when the bottom of all four feet of a machine are not on the same plane.
Spacer coupling	See "Floating Shaft Coupling."
Span	Center-to-center distance measured between two pulleys connected by a chain or belt.
Spiral gears or crossed-axis helical gears	When helical gears are used to connect nonparallel shafts
Split rigid coupling	A sleeve-type coupling that is split horizontally along the shaft and held together with bolts. Also known as a clamp coupling.
Sprocket	A tooth on the periphery of a wheel or cylinder that engages in the links of a chain.
Spur gear	Run together with other spur gears or parallel shafts, with internal gears on parallel shafts, and with a rack.
Static imbalance	Single-plane imbalance acting through the center of gravity of the rotor, perpendicular to the shaft axis. Also referred to as "Force Imbalance."
Stationary machine	The fixed machine component that is not adjusted during the alignment (the pump, fan, gear reducer, air compressor, etc. in most cases).
Stoddard solvent	A petroleum naphtha product with a comparatively narrow boiling range, used mostly for dry cleaning.
Strobe	A light pulse of very short duration.
Sump	A pit or tank.

Tempilstick	A crayon that, when applied to a surface, indicates when the surface temperature exceeds a given value by changing color.
Tensile strength	The maximum stress a material subjected to a stretching load can withstand without tearing.
Tension	The force exerted by a stretched object on a support.
Thermographic	Use of heat emissions of machinery or plant equipment as a monitoring and diagnostic predictive maintenance tool. For example, temperature differences on a coupling indicate misalignment and/or uneven mechanical forces.
Thrust	Axial forces and vibration created by the mechanical and/or dynamic operation of machine trains and/or process systems.
Timing belt	A positive drive belt that combines the advantages of belt drives with those of chains and gears. It has axial cogs molded on the underside of the belt, which fit into grooves on the pulley. The belt or chain prevents slip and makes accurate timing possible.
Tolerance	Generally referred to as manufacturing tolerance, a term that quantifies the allowable uncertainty in a dimension or allowable deviation from a baseline.
Tooth thickness	The distance along the pitch line or the circle form one side of a gear tooth to the other.
Torque	A moment/force couple applied to a rotor such as a shaft in order to sustain acceleration/load requirements. A twisting load imparted to shafts as the result of induced loads/speeds.

Torsion	A straining action produced by couples that act normal to the axis of a member, identified by a twisting deformation.
Transmission	System by which power is transmitted from the engine to a shaft or axle.
Trigonometric function	The real-valued functions such as sin(x), tan(x), and cos(x) obtained from studying certain ratios of the sides of a right triangle. Also known as circular functions.
Trigonometric leveling	A method of determining the difference of elevation between two points by using the principles of triangulation and trigonometric functions.
Two-way clutch	A clutch with a tooth profile (square or tapered) provides positive power transmission in both directions of rotation.
Vane	Casting on the backside of the impeller that reduces the pressure acting on the fluid behind the impeller.
V-belt	An endless V-shaped belt with a trapezoidal cross-section that runs in a pulley having a V-shaped groove.
Vibration	Continuous periodic change in a displacement with respect to a fixed reference.
Vibration/balance analyzer	A device that detects the level of imbalance, misalignment, etc., in a rotating part based on the measurement of vibration signals.
Viscosity	The resistance that a gaseous or liquid system offers to flow when it is subjected to a shear stress. Also known as flow resistance or internal friction.
Volute	Spiral part of a pump casing that accepts the fluid from the impeller and carries it to the outlet.

Wear	Deterioration of a surface due to material removal caused by relative motion between it and another part.
Whip	Phenomenon caused by pulsating loads in the drive system. Excessive belt whip is usually more common on long-center applications where the distance between the two pulley shafts is great.
Whip	Radial shaft movement. Also referred to as deflection.
Whole depth	The total height of a height of a tooth or the total depth of a tooth space.
Work	Transference of energy that occurs when a force is applied to a body that is moving in such a way that the force has a component in the direction of the body's motion.
Working depth	The depth of tooth engagement of two matching gears. It is the sum of their addendum.
Worm	A shank having at least one complete tooth (thread) around the pitch surface.
Worm gear	A gear with teeth cut on an angle to be driven by a worm; used to connect nonparallel, nonintersecting shafts.

INDEX

Printed and bound by CPI Group (UK) Ltd, Croydon, CR0 4YY

08/05/2025

01864816-0003